CAMBRIDGE LIBRARY COLLECTION

Books of enduring scholarly value

Mathematical Sciences

From its pre-historic roots in simple counting to the algorithms powering modern desktop computers, from the genius of Archimedes to the genius of Einstein, advances in mathematical understanding and numerical techniques have been directly responsible for creating the modern world as we know it. This series will provide a library of the most influential publications and writers on mathematics in its broadest sense. As such, it will show not only the deep roots from which modern science and technology have grown, but also the astonishing breadth of application of mathematical techniques in the humanities and social sciences, and in everyday life.

The Scientific Papers of Sir George Darwin

Sir George Darwin (1845-1912) was the second son and fifth child of Charles Darwin. After studying mathematics at Cambridge he read for the Bar, but soon returned to science and to Cambridge, where in 1883 he was appointed Plumian Professor of Astronomy and Experimental Philosophy. His work was concerned primarily with the effect of the sun and moon on tidal forces on Earth, and with the theoretical cosmogony which evolved from practical observation: he formulated the fission theory of the formation of the moon. This volume, published in 1916 after the author's death, includes a biographical memoir by his brother Sir Francis Darwin, his inaugural lecture and his lectures on George W. Hill's lunar theory.

Cambridge University Press has long been a pioneer in the reissuing of out-of-print titles from its own backlist, producing digital reprints of books that are still sought after by scholars and students but could not be reprinted economically using traditional technology. The Cambridge Library Collection extends this activity to a wider range of books which are still of importance to researchers and professionals, either for the source material they contain, or as landmarks in the history of their academic discipline.

Drawing from the world-renowned collections in the Cambridge University Library, and guided by the advice of experts in each subject area, Cambridge University Press is using state-of-the-art scanning machines in its own Printing House to capture the content of each book selected for inclusion. The files are processed to give a consistently clear, crisp image, and the books finished to the high quality standard for which the Press is recognised around the world. The latest print-on-demand technology ensures that the books will remain available indefinitely, and that orders for single or multiple copies can quickly be supplied.

The Cambridge Library Collection will bring back to life books of enduring scholarly value (including out-of-copyright works originally issued by other publishers) across a wide range of disciplines in the humanities and social sciences and in science and technology.

The Scientific Papers of Sir George Darwin

VOLUME 5:
SUPPLEMENTARY VOLUME

GEORGE HOWARD DARWIN
EDITED BY F.J.M. STRATTON
AND J. JACKSON

CAMBRIDGE
UNIVERSITY PRESS

CAMBRIDGE UNIVERSITY PRESS

Cambridge New York Melbourne Madrid Cape Town Singapore São Paolo Delhi

Published in the United States of America by Cambridge University Press, New York

www.cambridge.org
Information on this title: www.cambridge.org/9781108004480

© in this compilation Cambridge University Press 2009

This edition first published 1916
This digitally printed version 2009

ISBN 978-1-108-00448-0

SCIENTIFIC PAPERS

CAMBRIDGE UNIVERSITY PRESS

C. F. CLAY, Manager

London: FETTER LANE, E.C.

Edinburgh: 100 PRINCES STREET

New York: G. P. PUTNAM'S SONS

Bombay, Calcutta and Madras: MACMILLAN AND CO., Ltd.

Toronto: J. M. DENT AND SONS, Ltd.

Tokyo: THE MARUZEN-KABUSHIKI-KAISHA

From a water-colour drawing
by his daughter
Mrs Jacques Raverat

Emery Walker Ph.sc.

G. H. Darwin.

SCIENTIFIC PAPERS

BY

SIR GEORGE HOWARD DARWIN

K.C.B., F.R.S.

FELLOW OF TRINITY COLLEGE
PLUMIAN PROFESSOR IN THE UNIVERSITY OF CAMBRIDGE

VOLUME V
SUPPLEMENTARY VOLUME

CONTAINING

BIOGRAPHICAL MEMOIRS BY SIR FRANCIS DARWIN
AND PROFESSOR E. W. BROWN,
LECTURES ON HILL'S LUNAR THEORY, ETC.

EDITED BY

F. J. M. STRATTON, M.A., AND J. JACKSON, M.A., B.Sc.

Cambridge :
at the University Press
1916

𝕮𝖆𝖒𝖇𝖗𝖎𝖉𝖌𝖊:

PRINTED BY JOHN CLAY, M.A.

AT THE UNIVERSITY PRESS

PREFACE

BEFORE his death Sir George Darwin expressed the view that his lectures on Hill's Lunar Theory should be published. He made no claim to any originality in them, but he believed that a simple presentation of Hill's method, in which the analysis was cut short while the fundamental principles of the method were shewn, might be acceptable to students of astronomy. In this belief we heartily agree. The lectures might also with advantage engage the attention of other students of mathematics who have not the time to enter into a completely elaborated lunar theory. They explain the essential peculiarities of Hill's work and the method of approximation used by him in the discussion of an actual problem of nature of great interest. It is hoped that sufficient detail has been given to reveal completely the underlying principles, and at the same time not be too tedious for verification by the reader.

During the later years of his life Sir George Darwin collected his principal works into four volumes. It has been considered desirable to publish these lectures together with a few miscellaneous articles in a fifth volume of his works. Only one series of lectures is here given, although he lectured on a great variety of subjects connected with Dynamics, Cosmogony, Geodesy, Tides, Theories of Gravitation, etc. The substance of many of these is to be found in his scientific papers published in the four earlier volumes. The way in which in his lectures he attacked problems of great complexity by means of simple analytical methods is well illustrated in the series chosen for publication.

Two addresses are included in this volume. The one gives a view of the mathematical school at Cambridge about 1880, the other deals with the mathematical outlook of 1912.

The previous volumes contain all the scientific papers by Sir George Darwin published before 1910 which he wished to see reproduced. They do not include a large number of scientific reports on geodesy, the tides and other subjects which had involved a great deal of labour. Although the reports were of great value for the advancement and encouragement of science, he did not think it desirable to reprint them. We have not ventured to depart from his own considered decision; the collected lists at the beginning of these volumes give the necessary references for such papers as have been omitted. We are indebted to the Royal Astronomical Society for permission to complete Sir George Darwin's work on Periodic Orbits by reproducing his last published paper.

The opportunity has been taken of securing biographical memoirs of Darwin from two different points of view. His brother, Sir Francis Darwin, writes of his life apart from his scientific work, while Professor E. W. Brown, of Yale University, writes of Darwin the astronomer, mathematician and teacher.

F. J. M. S.
J. J.

GREENWICH,
   6 *December* 1915.

CONTENTS

MEMOIR OF SIR GEORGE DARWIN

BY

HIS BROTHER SIR FRANCIS DARWIN

George Howard, the fifth[1] child of Charles and Emma Darwin, was born at Down July 9th, 1845. Why he was christened[2] George, I cannot say. It was one of the facts on which we founded a theory that our parents lost their presence of mind at the font and gave us names for which there was neither the excuse of tradition nor of preference on their own part. His second name, however, commemorates his great-grandmother, Mary Howard, the first wife of Erasmus Darwin. It seems possible that George's ill-health and that of his father were inherited from the Howards. This at any rate was Francis Galton's view, who held that his own excellent health was a heritage from Erasmus Darwin's second wife. George's second name, Howard, has a certain appropriateness in his case for he was the genealogist and herald of our family, and it is through Mary Howard that the Darwins can, by an excessively devious route, claim descent from certain eminent people, e.g. John of Gaunt. This is shown in the pedigrees which George wrote out, and in the elaborate genealogical tree published in Professor Pearson's *Life of Francis Galton*. George's parents had moved to Down in September 1842, and he was born to those quiet surroundings of which Charles Darwin wrote "My life goes on like clock-work and I am fixed on the spot where I shall end it[3]." It would have been difficult to find a more retired place so near London. In 1842 a coach drive of some twenty miles was the only means of access to Down; and even now that railways have crept closer to it, it is singularly out of the world, with little to suggest the neighbourhood of London, unless it be the dull haze of smoke that sometimes clouds the sky. In 1842 such a village, communicating with the main lines of traffic only by stony tortuous lanes, may well have been enabled to retain something of its primitive character. Nor is it hard to believe in the smugglers and their strings of pack-horses making their way up from the lawless old villages of the Weald, of which the memory then still lingered.

[1] The third of those who survived childhood.
[2] At Maer, the Staffordshire home of his mother.
[3] *Life and Letters of Charles Darwin*, vol. I. p. 318.

George retained throughout life his deep love for Down. For the lawn with its bright strip of flowers; and for the row of big lime trees that bordered it. For the two yew trees between which we children had our swing, and for many another characteristic which had become as dear and as familiar to him as a human face. He retained his youthful love of the "Sand-walk," a little wood far enough from the house to have for us a romantic character of its own. It was here that our father took his daily exercise, and it has ever been haunted for us by the sound of his heavy walking stick striking the ground as he walked.

George loved the country round Down,—and all its dry chalky valleys of ploughed land with "shaws," i.e. broad straggling hedges on their crests, bordered by strips of flowery turf. The country is traversed by many foot-paths, these George knew well and used skilfully in our walks, in which he was generally the leader. His love for the house and the neighbourhood was I think entangled with his deepest feelings. In later years, his children came with their parents to Down, and they vividly remember his excited happiness, and how he enjoyed showing them his ancient haunts.

In this retired region we lived, as children, a singularly quiet life practically without friends and dependent on our brothers and sisters for companionship. George's earliest recollection was of drumming with his spoon and fork on the nursery table because dinner was late, while a barrel-organ played outside. Other memories were less personal, for instance the firing of guns when Sebastopol was supposed to have been taken. His diary of 1852 shows a characteristic interest in current events and in the picturesqueness of Natural History:

> The Duke is dead. Dodos are out of the world.

He perhaps carried rather far the good habit of re-reading one's favourite authors. He told his children that for a year or so he read through every day the story of Jack the Giant Killer, in a little chap-book with coloured pictures. He early showed signs of the energy which marked his character in later life. I am glad to remember that I became his companion and willing slave. There was much playing at soldiers, and I have a clear remembrance of our marching with toy guns and knapsacks across the field to the Sand-walk. There we made our bivouac with gingerbread, and milk, warmed (and generally smoked) over a "touch-wood" fire. I was a private while George was a sergeant, and it was part of my duty to stand sentry at the far end of the kitchen-garden until released by a bugle-call from the lawn. I have a vague remembrance of presenting my fixed bayonet at my father to ward off a kiss which seemed to me inconsistent with my military duties. Our imaginary names and heights were written up on the wall of the cloak-room. George, with romantic exactitude, made a small foot rule of such a size that he could conscientiously record his height as 6 feet and mine as slightly less, in accordance with my age and station.

Under my father's instruction George made spears with leaded heads which he hurled with remarkable skill by means of an Australian throwing stick. I used to skulk behind the big lime trees on the lawn in the character of victim, and I still remember the look of the spears flying through the air with a certain venomous waggle. Indoors, too, we threw at each other lead-weighted javelins which we received on beautiful shields made by the village carpenter and decorated with coats of arms.

Heraldry was a serious pursuit of his for many years, and the London Library copies of Guillim and Edmonson[1] were generally at Down. He retained a love of the science through life, and his copy of Percy's *Reliques* is decorated with coats of arms admirably drawn and painted. In later life he showed a power of neat and accurate draughtsmanship, and some of the illustrations in his father's books, e.g. in *Climbing Plants*, are by his hand.

His early education was given by governesses: but the boys of the family used to ride twice or thrice a week to be instructed in Latin by Mr Reed, the Rector of Hayes—the kindest of teachers. For myself, I chiefly remember the cake we used to have at 11 o'clock and the occasional diversion of looking at the pictures in the great Dutch bible. George must have impressed his parents with his solidity and self-reliance, since he was more than once allowed to undertake alone the 20 mile ride to the house of a relative at Hartfield in Sussex. For a boy of ten to bait his pony and order his luncheon at the Edenbridge inn was probably more alarming than the rest of the adventure. There is indeed a touch of David Copperfield in his recollections, as preserved in family tradition. "The waiter always said, 'What will you have for lunch, Sir?' to which he replied, 'What is there?' and the waiter said, 'Eggs and bacon'; and, though he hated bacon more than anything else in the world, he felt obliged to have it."

On August 16th, 1856, George was sent to school. Our elder brother, William, was at Rugby, and his parents felt his long absences from home such an evil that they fixed on the Clapham Grammar School for their younger sons. Besides its nearness to Down, Clapham had the merit of giving more mathematics and science than could then be found in public schools. It was kept by the Rev. Charles Pritchard[2], a man of strong character and with a gift for teaching mathematics by which George undoubtedly profited. In (I think) 1861 Pritchard left Clapham and was succeeded by the Rev. Alfred Wrigley, a man of kindly mood but without the force or vigour of Pritchard. As a mathematical instructor I imagine

---

[1] Guillim, John, *A display of heraldry*, 6th ed., folio 1724. Edmonson, J., *A complete body of heraldry*, folio 1780.

[2] Afterwards Savilian Professor of Astronomy at Oxford. Born 1808, died 1893.

Wrigley was a good drill-master rather than an inspiring teacher. Under him the place degenerated to some extent; it no longer sent so many boys to the Universities, and became more like a "crammer's" and less like a public school. My own recollections of George at Clapham are coloured by an abiding gratitude for his kindly protection of me as a shrinking and very unhappy "new boy" in 1860.

George records in his diary that in 1863 he tried in vain for a Minor Scholarship at St John's College, Cambridge, and again failed to get one at Trinity in 1864, though he became a Foundation Scholar in 1866. These facts suggested to me that his capacity as a mathematician was the result of slow growth. I accordingly applied to Lord Moulton, who was kind enough to give me his impressions:

> My memories of your brother during his undergraduate career correspond closely to your suggestion that his mathematical power developed somewhat slowly and late. Throughout most if not the whole of his undergraduate years he was in the same class as myself and Christie, the ex-Astronomer Royal, at Routh's[1]. We all recognised him as one who was certain of being high in the Tripos, but he did not display any of that colossal power of work and taking infinite trouble that characterised him afterwards. On the contrary, he treated his work rather jauntily. At that time his health was excellent and he took his studies lightly so that they did not interfere with his enjoyment of other things[2]. I remember that as the time of the examination came near I used to tell him that he was unfairly handicapped in being in such robust health and such excellent spirits.
>
> Even when he had taken his degree I do not think he realised his innate mathematical power....It has been a standing wonder to me that he developed the patience for making the laborious numerical calculations on which so much of his most original work was necessarily based. He certainly showed no tendency in that direction during his undergraduate years. Indeed he told me more than once in later life that he detested Arithmetic and that these calculations were as tedious and painful to him as they would have been to any other man, but that he realised that they must be done and that it was impossible to train anyone else to do them.

As a Freshman he "kept" (i.e. lived) in A 6, the staircase at the N.W. corner of the New Court, afterwards moving to F 3 in the Old Court, pleasant rooms entered by a spiral staircase on the right of the Great Gate. Below him, in the ground floor room, now used as the College offices, lived Mr Colvill, who remained a faithful but rarely seen friend as long as George lived.

Lord Moulton, who, as we have seen, was a fellow pupil of George's at Routh's, was held even as a Freshman to be an assured Senior Wrangler,

---

[1] The late Mr Routh was the most celebrated Mathematical "Coach" of his day.

[2] Compare Charles Darwin's words: "George has not slaved himself, which makes his success the more satisfactory" (*More Letters of C. Darwin*, vol. II. p. 287).

a prophecy that he easily made good.   The second place was held by George, and was a much more glorious position than he had dared to hope for.   In those days the examiners read out the list in the Senate House at an early hour, 8 a.m. I think.   George remained in bed and sent me to bring the news.   I remember charging out through the crowd the moment the magnificent "Darwin of Trinity" had followed the expected "Moulton of St John's." I have a general impression of a cheerful crowd sitting on George's bed and literally almost smothering him with congratulations.   He received the following characteristic letter from his father[1]:

Down, *Jan. 24th* [1868].

My dear old fellow,

      I am so pleased.   I congratulate you with all my heart and soul.   I always said from your early days that such energy, perseverance and talent as yours would be sure to succeed: but I never expected such brilliant success as this.   Again and again I congratulate you.   But you have made my hand tremble so I can hardly write.   The telegram came here at eleven.   We have written to W. and the boys.

    God bless you, my dear old fellow—may your life so continue.

Your affectionate Father,

CH. DARWIN.

In those days the Tripos examination was held in the winter, and the successful candidates got their degrees early in the Lent Term: George records in his diary that he took his B.A. on January 25th, 1868: also that he won the second of the two Smith's Prizes,—the first being the natural heritage of the Senior Wrangler.   There is little to record in this year.   He had a pleasant time in the summer coaching Clement Bunbury, the nephew of Sir Charles, at his beautiful place Barton Hall in Suffolk. In the autumn he was elected a Fellow of Trinity, as he records, "with Galabin, young Niven, Clifford, [Sir Frederick] Pollock, and [Sir Sidney] Colvin."   W. K. Clifford was the well-known brilliant mathematician who died comparatively early.

Chief among his Cambridge friends were the brothers Arthur, Gerald and Frank Balfour.   The last-named was killed, aged 31, in a climbing accident in 1882 on the Aiguille Blanche near Courmayeur.   He was remarkable both for his scientific work and for his striking and most lovable personality.   George's affection for him never faded.   Madame Raverat remembers her father (not long before his death) saying with emotion, "I dreamed Frank Balfour was alive."   I imagine that tennis was the means of bringing George into contact with Mr Arthur Balfour.   What began in this chance way grew into an enduring friendship, and George's diary shows how much kindness and hospitality he received from Mr Balfour.   George had also the

[1] *Emma Darwin, A Century of Family Letters*, vol. II. p. 186.

advantage of knowing Lord Rayleigh at Cambridge, and retained his friend-ship through his life.

In the spring of 1869 he was in Paris for two months working at French. His teacher used to make him write original compositions, and George gained a reputation for humour by giving French versions of all the old Joe Millers and ancient stories he could remember.

It was his intention to make the Bar his profession[1], and in October 1869 we find him reading with Mr Tatham, in 1870 and 1872 with the late Mr Montague Crackenthorpe (then Cookson). Again, in November 1871, he was a pupil of Mr W. G. Harrison. The most valued result of his legal work was the friendship of Mr and Mrs Crackenthorpe, which he retained throughout his life. During these years we find the first indications of the circumstances which forced him to give up a legal career—namely, his failing health and his growing inclination towards science[2]. Thus in the summer of 1869, when we were all at Caerdeon in the Barmouth valley, he writes that he "fell ill": and again in the winter of 1871. His health deteriorated markedly during 1872 and 1873. In the former year he went to Malvern and to Homburg without deriving any advantage. I have an impression that he did not expect to survive these attacks; but I cannot say at what date he made this forecast of an early death. In January 1873 he tried Cannes: and "came back very ill." It was in the spring of this year that he first consulted Dr (afterwards Sir Andrew) Clark, from whom he received the kindest care. George suffered from digestive troubles, sickness and general discomfort and weakness. Dr Clark's care probably did what was possible to make life more bearable, and as time went on his health gradually improved. In 1894 he consulted the late Dr Eccles, and by means of the rest-cure, then something of a novelty, his weight increased from 9 stone to 9 stone 11 pounds. I gain the impression that this treatment produced a permanent improvement, although his health remained a serious handicap throughout his life.

Meanwhile he had determined on giving up the Bar, and settled, in October 1873, when he was 28 years old, at Trinity in Nevile's Court next the Library (G 4). His diary continues to contain records of ill-health and of various holidays in search of improvement. Thus in 1873 we read "Very bad during January. Went to Cannes and stayed till the end of April." Again in 1874, "February to July very ill." In spite of unwellness he began in 1872—3 to write on various subjects. He sent to *Macmillan's Magazine*[3] an enter-taining article, "Development in Dress," where the various survivals in modern

---

[1] He was called in 1874 but did not practise.

[2] As a boy he had energetically collected Lepidoptera during the years 1858—61, but the first vague indications of a leaning towards physical science may perhaps be found in his joining the Sicilian eclipse expedition, Dec. 1870—Jan. 1871. It appears from *Nature*, Dec. 1, 1870, that George was told off to make sketches of the Corona.

[3] *Macmillan's Magazine*, 1872, vol. xxvi. pp. 410—416.

costume were recorded and discussed from the standpoint of evolution.   In
1873 he wrote " On beneficial restriction to liberty of marriage[1]," a eugenic
article for which he was attacked with gross unfairness and bitterness by the
late St George Mivart.   He was defended by Huxley, and Charles Darwin
formally ceased all intercourse with Mivart.   We find mention of a " Globe
Paper for the British Association" in 1873.   And in the following year he
read a contribution on "Probable Error" to the Mathematical Society[2]—on
which he writes in his diary, " found it was old."   Besides another paper in the
Messenger of Mathematics, he reviewed " Whitney on Language[3]," and wrote
a " defence of Jevons" which I have not been able to trace.   In 1875 he
was at work on the " flow of pitch," on an " equipotential tracer," on slide
rules, and sent a paper on " Cousin Marriages " to the Statistical Society[4].   It
is not my province to deal with these papers; they are here of interest as
showing his activity of mind and his varied interests, features in character
which were notable throughout his life.

The most interesting entry in his diary for 1875 is " Paper on Equi-
potentials much approved by Sir W. Thomson."   This is the first notice of an
association of primary importance in George's scientific career.   Then came
his memoir " On the influence of geological changes in the earth's axis of
rotation."   Lord Kelvin was one of the referees appointed by the Council of
the Royal Society to report on this paper, which was published in the *Philosophical Transactions* in 1877.

In his diary, November 1878, George records " paper on tides ordered to
be printed."   This refers to his work " On the bodily tides of viscous and
semi-elastic spheroids, etc.," published in the *Phil. Trans.* in 1879.   It was in
regard to this paper that his father wrote to George on October 29th, 1878[5]:

> My dear old George,
>   I have been quite delighted with your letter and read it all
> with eagerness.   You were very good to write it.   All of us are
> delighted, for considering what a man Sir William Thomson is, it is
> most grand that you should have staggered him so quickly, and that he
> should speak of your 'discovery, etc.'...Hurrah for the bowels of the
> earth and their viscosity and for the moon and for the Heavenly bodies
> and for my son George (F.R.S. very soon)...[6].

The bond of pupil and master between George Darwin and Lord Kelvin,
originating in the years 1877—8, was to be a permanent one, and developed

---

[1] *Contemporary Review*, 1873, vol. xxii. pp. 412—426.

[2] Not published.

[3] *Contemporary Review*, 1874, vol. xxiv. pp. 894—904.

[4] *Journal of the Statistical Society*, 1875, vol. xxxviii. pt 2, pp. 153—182, also pp. 183—184,
and pp. 344—348.

[5] Probably he heard informally at the end of October what was not formally determined till
November.

[6] *Emma Darwin, A Century of Family Letters*, 1915, vol. ii. p. 233.

not merely into scientific co-operation but into a close friendship. Sir Joseph Larmor has recorded[1] that George's "tribute to Lord Kelvin, to whom he dedicated volume I of his Collected Papers[2]...gave lively pleasure to his master and colleague." His words were:

> Early in my scientific career it was my good fortune to be brought into close personal relationship with Lord Kelvin. Many visits to Glasgow and to Largs have brought me to look up to him as my master, and I cannot find words to express how much I owe to his friendship and to his inspiration.

During these years there is evidence that he continued to enjoy the friendship of Lord Rayleigh and of Mr Balfour. We find in his diary records of visits to Terling and to Whittingehame, or of luncheons at Mr Balfour's house in Carlton Gardens for which George's scientific committee work in London gave frequent opportunity. In the same way we find many records of visits to Francis Galton, with whom he was united alike by kinship and affection.

Few people indeed can have taken more pains to cultivate friendship than did George. This trait was the product of his affectionate and eminently sociable nature and of the energy and activity which were his chief characteristics. In earlier life he travelled a good deal in search of health[3], and in after years he attended numerous congresses as a representative of scientific bodies. He thus had unusual opportunities of making the acquaintance of men of other nationalities, and some of his warmest friendships were with foreigners. In passing through Paris he rarely failed to visit M. and Mme d'Estournelles and "the d'Abbadies." It was in Algiers in 1878 and 1879 that he cemented his friendship with the late J. F. MacLennan, author of *Primitive Marriage*; and in 1880 he was at Davos with the same friends. In 1881 he went to Madeira, where he received much kindness from the Blandy family—doubtless through the recommendation of Lady Kelvin.

Cambridge.

We have seen that George was elected a Fellow of Trinity in October 1868, and that five years later (Oct. 1873) he began his second lease of a Cambridge existence. There is at first little to record: he held at this time no official position, and when his Fellowship expired he continued to live in College busy with his research work and laying down the earlier tiers

---

[1] *Nature*, Dec. 12, 1912.

[2] It was in 1907 that the Syndics of the Cambridge University Press asked George to prepare a reprint of his scientific papers, which the present volume brings to an end. George was deeply gratified at an honour that placed him in the same class as Lord Kelvin, Stokes, Cayley, Adams, Clerk Maxwell, Lord Rayleigh and other men of distinction.

[3] Thus in 1872 he was in Homburg, 1873 in Cannes, 1874 in Holland, Belgium, Switzerland and Malta, 1876 in Italy and Sicily.

of the monumental series of papers in the present volumes. This soon led to his being proposed (in Nov. 1877) for the Royal Society, and elected in June 1879. The principal event in this stage of his Cambridge life was his election[1] in 1883 as Plumian Professor of Astronomy and Experimental Philosophy. His predecessor in the Chair was Professor Challis, who had held office since 1836, and is now chiefly remembered in connection with Adams and the planet Neptune. The professorship is not necessarily connected with the Observatory, and practical astronomy formed no part of George's duties. His lectures being on advanced mathematics usually attracted but few students; in the Long Vacation however, when he habitually gave one of his courses, there was often a fairly large class.

George's relations with his class have been sympathetically treated by Professor E. W. Brown, than whom no one can speak with more authority, since he was one of my brother's favourite pupils.

In the late '70's George began to be appointed to various University Boards and Syndicates. Thus from 1878—82 he was on the Museums and Lecture Rooms Syndicate. In 1879 he was placed on the Observatory Syndicate, of which he became an official member in 1883 on his election to the Plumian Professorship. In the same way he was on the Special Board for Mathematics. He was on the Financial Board from 1900—1 to 1903—4 and on the Council of the Senate in 1905—6 and 1908—9. But he never became a professional syndic—one of those virtuous persons who spend their lives in University affairs. In his obituary of George (*Nature*, Dec. 12, 1912), Sir Joseph Larmor writes:

> In the affairs of the University of which he was an ornament, Sir George Darwin made a substantial mark, though it cannot be said that he possessed the patience in discussion that is sometimes a necessary condition to taking a share in its administration. But his wide acquaintance and friendships among the statesmen and men of affairs of the time, dating often from undergraduate days, gave him openings for usefulness on a wider plane. Thus, at a time when residents were bewailing even more than usual the inadequacy of the resources of the University for the great expansion which the scientific progress of the age demanded, it was largely on his initiative that, by a departure from all precedent, an unofficial body was constituted in 1899 under the name

---

[1] The voting at University elections is in theory strictly confidential, but in practice this is unfortunately not always the case. George records in his diary the names of the five who voted for him and of the four who supported another candidate. None of the electors are now living. The election occurred in January, and in June he had the great pleasure and honour of being re-elected to a Trinity Fellowship. His daughter, Madame Raverat, writes: "Once, when I was walking with my father on the road to Madingley village, he told me how he had walked there, on the first Sunday he ever was at Cambridge, with two or three other freshmen; and how, when they were about opposite the old chalk pit, one of them betted him £20 that he (my father) would never be a professor of Cambridge University: and said my father, with great indignation, 'He never paid me.'"

of the Cambridge University Association, to promote the further endowment of the University by interesting its graduates throughout the Empire in its progress and its more pressing needs. This important body, which was organised under the strong lead of the late Duke of Devonshire, then Chancellor, comprises as active members most of the public men who owe allegiance to Cambridge, and has already by its interest and help powerfully stimulated the expansion of the University into new fields of national work; though it has not yet achieved financial support on anything like the scale to which American seats of learning are accustomed.

The Master of Christ's writes:

May 31st, 1915.

My impression is that George did not take very much interest in the petty details which are so beloved by a certain type of University authority. 'Comma hunting' and such things were not to his taste, and at Meetings he was often rather distrait: but when anything of real importance came up he was of extraordinary use. He was especially good at drafting letters, and over anything he thought promoted the advancement of the University along the right lines he would take endless trouble—writing and re-writing reports and letters till he got them to his taste. The sort of movements which interested him most were those which connected Cambridge with the outside world. He was especially interested in the Appointments Board. A good many of us constantly sought his advice and nearly always took it: but, as I say, I do not think he cared much about the 'parish pump,' and was usually worried at long Meetings.

Professor Newall has also been good enough to give me his impressions:

His weight in the Committees on which I have had personal experience of his influence seems to me to have depended in large measure on his realising very clearly the distinction between the importance of ends to be aimed at and the difficulty of harmonising the personal characteristics of the men who might be involved in the work needed to attain the ends. The ends he always took seriously; the crotchets he often took humorously, to the great easement of many situations that are liable to arise on a Committee. I can imagine that to those who had corns his direct progress may at times have seemed unsympathetic and hasty. He was ready to take much trouble in formulating statements of business with great precision—a result doubtless of his early legal experiences. I recall how he would say, 'If a thing has to be done, the minute should if possible make some individual responsible for doing it.' He would ask, 'Who is going to do the work? If a man has to take the responsibility, we must do what we can to help him and not hamper him by unnecessary restrictions and criticisms.' His helpfulness came from his quickness in seizing the important point and his readiness to take endless trouble in the important work of looking into details before and after the meetings. The amount of work that he did in response to the requirements of various Committees was very great, and it was curious to realise in how many cases he seemed to have diffidence as to the value of his contributions.

But on the whole the work which, in spite of ill-health, he was able to carry out in addition to professional duties and research, was given to matters unconnected with the University, but of a more general importance. To these we shall return.

In 1884 he became engaged to Miss Maud Du Puy of Philadelphia. She came of an old Huguenot stock, descending from Dr John Du Puy who was born in France in 1679 and settled in New York in 1713. They were married on July 22nd, 1884, and this event happily coloured the remainder of George's life. As time went on and existence became fuller and busier, she was able by her never-failing devotion to spare him much arrangement and to shield him from fatigue and anxiety. In this way he was helped and protected in the various semi-public functions in which he took a principal part. Nor was her help valued only on these occasions, for indeed the comfort and happiness of every day was in her charge. There is a charming letter[1] from George's mother, dated April 15th, 1884:

> Maud had to put on her wedding-dress in order to say at the Custom-house in America that she had worn it, so we asked her to come down and show it to us. She came down with great simplicity and quietness...only really pleased at its being admired and at looking pretty herself, which was strikingly the case. She was a little shy at coming in, and sent in Mrs Jebb to ask George to come out and see it first and bring her in. It was handsome and simple. I like seeing George so frivolous, so deeply interested in which diamond trinket should be my present, and in her new Paris morning dress, in which he felt quite unfit to walk with her.

Later, probably in June, George's mother wrote[2] to Miss Du Puy, "Your visit here was a great happiness to me, as something in you (I don't know what) made me feel sure you would always be sweet and kind to George when he is ill and uncomfortable." These simple and touching words may be taken as a forecast of his happy married life.

In March 1885 George acquired by purchase the house Newnham Grange[3], which remained his home to the end of his life. It stands at the southern end of the Backs, within a few yards of the river where it bends eastward in flowing from the upper to the lower of the two Newnham water-mills. I remember forebodings as to dampness, but they proved wrong— even the cellars being remarkably dry. The house is built of faded yellowish bricks with old tiles on the roof, and has a pleasant home-like air.

---

[1] *Emma Darwin, A Century of Family Letters.* Privately printed, 1904, vol. II. p. 350.

[2] *Emma Darwin, A Century of Family Letters*, 1912, vol. II. p. 266.

[3] At that time it was known simply as *Newnham*, but as this is the name of the College and was also in use for a growing region of houses, the Darwins christened it Newnham Grange. The name Newnham is now officially applied to the region extending from Silver Street Bridge to the Barton Road.

It was formerly the house of the Beales family[1], one of the old merchant stocks of Cambridge. This fact accounts for the great barn-like granaries which occupied much of the plot near the high road. These buildings were in part pulled down, thus making room for a lawn tennis court, while what was not demolished made a gallery looking on the court as well as play-room for the children. At the eastern end of the property a cottage and part of the granaries were converted into a small house of an attractively individual character, for which I think tenants have hitherto been easily found among personal friends. It is at present inhabited by Lady Corbett. One of the most pleasant features of the Grange was the flower-garden and rockery on the other side of the river, reached by a wooden bridge and called "the Little Island[2]." The house is conveniently close to the town, yet has a most pleasant outlook, to the north over the Backs while there is the river and the Fen to the south. The children had a den or house in the branches of a large copper beech tree, overhanging the river. They were allowed to use the boat, which was known as the *Griffin* from the family crest with which it was adorned. None of them were drowned, though accidents were not unknown; in one of these an eminent lady and well-known writer, who was inveigled on to the river by the children, had to wade to shore near Silver Street bridge owing to the boat running aground.

The Darwins had five children, of whom one died an infant: of the others, Charles Galton Darwin has inherited much of his father's mathematical ability, and has been elected to a Mathematical Lectureship at Christ's College. He is now in the railway service of the Army in France. The younger son, William, has a commission in the 18th Battalion of the Durham Light Infantry. George's elder daughter is married to Monsieur Jacques Raverat. Her skill as an artist has perhaps its hereditary root in her father's draughtsmanship. The younger daughter Margaret lives with her mother.

George's relations with his family were most happy. His diary never fails to record the dates on which the children came home, or the black days which took them back to school. There are constantly recurring entries in his diary of visits to the boys at Marlborough or Winchester. Or of the

---

[1] The following account of Newnham Grange is taken from C. H. Cooper's *Memorials of Cambridge*, 1866, vol. III. p. 262 (note) :—"The site of the hermitage was leased by the Corporation to Oliver Grene, 20 Sep., 31 Eliz. [1589]. It was in 1790 leased for a long term to Patrick Beales, from whom it came to his brother S. P. Beales, Esq., who erected thereon a substantial mansion and mercantile premises now occupied by his son Patrick Beales, Esq., alderman, who purchased the reversion from the Corporation in 1839." Silver Street was formerly known as Little Bridges Street, and the bridges which gave it this name were in charge of a hermit, hence the above reference to the hermitage.

[2] This was to distinguish it from the "Big Island," both being leased from the town. Later George acquired in the same way the small oblong kitchen garden on the river bank, and bought the freehold of the Lammas land on the opposite bank of the river.

journeys to arrange for the schooling of the girls in England or abroad. The parents took pains that their children should have opportunities of learning conversational French and German.

George's characteristic energy showed itself not only in these ways but also in devising bicycling expeditions and informal picnics, for the whole family, to the Fleam Dyke, to Whittlesford, or other pleasant spots near home —and these excursions he enjoyed as much as anyone of the party. As he always wished to have his children with him, one or more generally accompanied him and his wife when they attended congresses or other scientific gatherings abroad.

His house was the scene of many Christmas dinners, the first of which I find any record being in 1886. These meetings were often made an occasion for plays acted by the children; of these the most celebrated was a Cambridge version of *Romeo and Juliet*, in which the hero and heroine were scions of the rival factions of Trinity and St John's.

Games and Pastimes.

As an undergraduate George played tennis—not the modern out-door game, but that regal pursuit which is sometimes known as the game of kings and otherwise as the king of games. When George came up as an undergraduate there were two tennis courts in Cambridge, one in the East Road, the other being the ancient one that gave its name to Tennis Court Road and was pulled down to make room for the new buildings of Pembroke. In this way was destroyed the last of the College tennis courts of which we read in Mr Clark's *History*. I think George must have had pleasure in the obvious development of the tennis court from some primaeval court-yard in which the *pent-house* was the roof of a shed, and the *grille* a real window or half-door. To one brought up on evolution there is also a satisfaction about the French terminology which survives in e.g. the *Tambour* and the *Dedans*. George put much thought into acquiring a correct style of play—for in tennis there is a religion of attitude corresponding to that which painfully regulates the life of the golfer. He became a good tennis player as an undergraduate, and was in the running for a place in the inter-University match. The marker at the Pembroke court was Henry Harradine, whom we all sincerely liked and respected, but he was not a good teacher, and it was only when George came under Henry's sons, John and Jim Harradine, at the Trinity and Clare courts, that his game began to improve. He continued to play tennis for some years, and only gave it up after a blow from a tennis ball in January 1895 had almost destroyed the sight of his left eye.

In 1910 he took up archery, and zealously set himself to acquire the correct mode of standing, the position of the head and hands, etc. He kept an archery diary in which each day's shooting is carefully analysed and the

results given in percentages. In 1911 he shot on 131 days: the last occasion
on which he took out his bow was September 13, 1912.

I am indebted to Mr H. Sherlock, who often shot with him at Cambridge,
for his impressions. He writes: "I shot a good deal with your brother the
year before his death; he was very keen on the sport, methodical and pains-
taking, and paid great attention to style, and as he had a good natural
'loose,' which is very difficult to acquire, there is little doubt (notwithstanding
that he came to Archery rather late in life) that had he lived he would have
been above the average of the men who shoot fairly regularly at the public
Meetings." After my brother's death, Mr Sherlock was good enough to look
at George's archery note-book. "I then saw," he writes, "that he had
analysed them in a way which, so far as I am aware, had never been done
before." Mr Sherlock has given examples of the method in a sympathetic
obituary published (p. 273) in *The Archer's Register*[1]. George's point was
that the traditional method of scoring is not fair in regard to the areas of the
coloured rings of the target. Mr Sherlock records in his *Notice* that George
joined the Royal Toxophilite Society in 1912, and occasionally shot in the
Regent's Park. He won the Norton Cup and Medal (144 arrows at 120
yards) in 1912.

There was a billiard table at Down, and George learned to play fairly
well though he had no pretension to real proficiency. He used to play at
the Athenaeum, and in 1911 we find him playing there in the Billiard
Handicap, but a week later he records in his diary that he was "knocked
out."

Scientific Committees.

George served for many years on the Solar Physics Committee and on
the Meteorological Council. With regard to the latter, Sir Napier Shaw
has at my request supplied the following note :—

It was in February 1885 upon the retirement of Warren De la Rue
that your brother George, by appointment of the Royal Society, joined
the governing body of the Meteorological Office, at that time the
Meteorological Council. He remained a member until the end of the
Council in 1905 and thereafter, until his death, he was one of the two
nominees of the Royal Society upon the Meteorological Committee, the
new body which was appointed by the Treasury to take over the control
of the administration of the Office.

It will be best to devote a few lines to recapitulating the salient
features of the history of the official meteorological organisation because,
otherwise, it will be difficult for anyone to appreciate the position in
which Darwin was placed.

---

[1] *The Archer's Register* for 1912—1913, by H. Walrond.   London, *The Field* Office, 1913.

In 1854 a department of the Board of Trade was constituted under Admiral R. FitzRoy to collect and discuss meteorological information from ships, and in 1860, impressed by the loss of the 'Royal Charter,' FitzRoy began to collect meteorological observations by telegraph from land stations and chart them. Looking at a synchronous chart and conscious that he could gather from it a much better notion of coming weather than anyone who had only his own visible sky and barometer to rely upon, he formulated 'forecasts' which were published in the newspapers and 'storm warnings' which were telegraphed to the ports.

This mode of procedure, however tempting it might be to the practical man with the map before him, was criticised as not complying with the recognised canons of scientific research, and on FitzRoy's untimely death in 1865 the Admiralty, the Board of Trade and the Royal Society elaborated a scheme for an office for the study of weather in due form under a Director and Committee, appointed by the Royal Society, and they obtained a grant in aid of £10,000 for this purpose. In this transformation it was Galton, I believe, who took a leading part and to him was probably due the initiation of the new method of study which was to bring the daily experience, as represented by the map, into relation with the continuous records of the meteorological elements obtained at eight observatories of the Kew type, seven of which were immediately set on foot, and Galton devoted an immense amount of time and skill to the reproduction of the original curves so that the whole sequence of phenomena at the seven observatories could be taken in at a glance. Meanwhile the study of maps was continued and a good deal of progress was made in our knowledge of the laws of weather.

But in spite of the wealth of information the generalisations were empirical and it was felt that something more than the careful examination of records was required to bring the phenomena of weather within the rule of mathematics and physics, so in 1876 the constitution of the Office was changed and the direction of its work was placed in Commission with an increased grant. The Commissioners, collectively known as the Meteorological Council, were a remarkably distinguished body of fellows of the Royal Society, and when Darwin took the place of De la Rue, the members were men subsequently famous, as Sir Richard Strachey, Sir William Wharton, Sir George Stokes, Sir Francis Galton, Sir George Darwin, with E. J. Stone, a former Astronomer Royal for the Cape.

It was understood that the attack had to be made by new methods and was to be entrusted partly to members of the Council themselves, with the staff of the Office behind them, and partly to others outside who should undertake researches on special points. Sir Andrew Noble, Sir William Abney, Dr W. J. Russell, Mr W. H. Dines, your brother Horace and myself came into connection with the Council in this way.

Two important lines of attack were opened up within the Council itself. The first was an attempt, under the influence of Lord Kelvin, to base an explanation of the sequence of weather upon harmonic analysis. As the phenomena of tides at any port could be synthesized by the combinations of waves of suitable period and amplitude, so the sequence of weather could be analysed into constituent oscillations the general relations of which would be recognisable although the original composite result was intractable on direct inspection. It was while this enterprise was in progress that Darwin was appointed to the Council. His experience with tides and tidal analysis was in a way his title to admission. He and Stokes were the mathematicians of the Council and were looked to for expert guidance in the undertaking. At first the individual curves were submitted to analysis in a harmonic analyser specially built for the purpose, the like of which Darwin had himself used or was using for his work on tides; but afterwards it was decided to work arithmetically with the numbers derived from the tabulation of the curves; and the identity of the individual curves was merged in 'five-day means.' The features of the automatic records from which so much was hoped in 1865, after twelve years of publication in facsimile, were practically never seen outside the room in the Office in which they were tabulated.

It is difficult at this time to point to any general advances in meteorology which can be attributed to the harmonic analyser or its arithmetical equivalent as a process of discussion, though it still remains a powerful method of analysis. It has, no doubt, helped towards the recognition of the ubiquity and simultaneity of the twelve-hour term in the diurnal change of pressure which has taken its place among fundamental generalisations of meteorology and the curious double diurnal change in the wind at any station belongs to the same category; but neither appears to have much to do with the control of weather. Probably the real explanation of the comparative fruitlessness of the effort lies in the fact that its application was necessarily restricted to the small area of the British Isles instead of being extended, in some way or other, to the globe.

It is not within my recollection that Darwin was particularly enthusiastic about the application of harmonic analysis. When I was appointed to the Council in 1897, the active pursuit of the enterprise had ceased. Strachey who had taken an active part in the discussion of the results and contributed a paper on them to the Philosophical Transactions, was still hopeful of basing important conclusions upon the seasonal peculiarities of the third component, but the interest of other members of the Council was at best languid.

The other line of attack was in connection with synoptic charts. For the year from August 1892 to August 1893 there was an international scheme for circumpolar observations in the Northern Hemisphere, and in connection therewith the Council undertook the preparation of daily synoptic charts of the Atlantic and adjacent land areas. A magnificent series of charts was produced and published from which great results were anticipated. But again the conclusions drawn from cursory inspection were disappointing. At that time the suggestion that weather travelled across the Atlantic in so orderly a manner that our weather could be notified four or five days in advance from New York had a considerable vogue and the facts disclosed by the charts put an end to any hope of the practical development of that suggestion. Darwin was very active in endeavouring to obtain the help of an expert in physics for the discussion of the charts from a new point of view, but he was unsuccessful.

Observations at High Level Stations were also included in the

Council's programme.   A station was maintained at Hawes Junction for some years, and the Observatories on Ben Nevis received their support.   But when I joined the Council in 1897 there was a pervading sense of discouragement.   The forecasting had been restored as the result of the empirical generalisations based on the work of the years 1867 to 1878, but the study had no attractions for the powerful analytical minds of the Council; and the work of the Office had settled down into the assiduous compilation of observations from sea and land and the regular issue of forecasts and warnings in the accustomed form.   The only part which I can find assigned to Darwin with regard to forecasting is an endeavour to get the forecast worded so as not to suggest more assurance than was felt.

I do not think that Darwin addressed himself spontaneously to meteorological problems, but he was always ready to help.   He was very regular in his attendance at Council and the Minutes show that after Stokes retired all questions involving physical measurement or mathematical reasoning were referred to him.   There is a short and very characteristic report from him on the work of the harmonic analyser and a considerable number upon researches by Mr Dines or Sir G. Stokes on anemometers.   It is hardly possible to exaggerate his aptitude for work of that kind.   He could take a real interest in things that were not his own.   He was full of sympathy and appreciation for efforts of all kinds, especially those of young men, and at the same time, using his wide experience, he was perfectly frank and fearless not only in his judgment but also in the expression of it.   He gave one the impression of just protecting himself from boredom by habitual loyalty and a finely tempered sense of duty.   My earliest recollection of him on the Council is the thrilling production of a new version of the Annual Report of the Council which he had written because the original had become more completely 'scissors and paste' than he could endure.

After the Office came into my charge in 1900, so long as he lived, I never thought of taking any serious step without first consulting him and he was always willing to help by his advice, by his personal influence and by his special knowledge.   For the first six years of the time I held a college fellowship with the peculiar condition of four public lectures in the University each year and no emolument.   One year, when I was rather overdone, Darwin took the course for me and devoted the lectures to Dynamical Meteorology.   I believe he got it up for the occasion, for he professed the utmost diffidence about it, but the progress which we have made in recent years in that subject dates from those lectures and the correspondence which arose upon them.

In Council it was the established practice to proceed by agreement and not by voting; he had a wonderful way of bringing a discussion to a head by courageously 'voicing' the conclusion to which it led and frankly expressing the general opinion without hurting anybody's feelings.

This letter has, I fear, run to a great length, but it is not easy to give expression to the powerful influence which he exercised upon all departments of official meteorology without making formal contributions to meteorological literature.   He gave me a note on a curious point in the evaluation of the velocity equivalents of the Beaufort Scale which is published in the Office Memoirs No. 180, and that is all I have to show in print, but he was in and behind everything that was done and personally, I need hardly add, I owe to him much more than this or any other letter can fully express.

On May 6, 1904, he was elected President of the British Association —the South African meeting.

On July 29, 1905, he embarked with his wife and his son Charles and arrived on August 15 at the Cape, where he gave the first part of his Presidential Address.   Here he had the pleasure of finding as Governor Sir Walter Hely-Hutchinson, whom he had known as a Trinity undergraduate. He was the guest of the late Sir David Gill, who remained a close friend for the rest of his life.   George's diary gives his itinerary—which shows the trying amount of travel that he went through.   A sample may be quoted:

August 19   Embark,
,,   22   Arrive at Durban,
,,   23   Mount Edgecombe,
,,   24   Pietermaritzburg,
,,   26   Colenso,
,,   27   Ladysmith,
,,   28   Johannesburg.

At Johannesburg he gave the second half of his Address.   Then on by Bloemfontein, Kimberley, Bulawayo, to the Victoria Falls, where a bridge had to be opened.   Then to Portuguese Africa on September 16, 17, where he made speeches in French and English.   Finally he arrived at Suez on October 4 and got home October 18.

It was generally agreed that his Presidentship was a conspicuous success. The following appreciation is from the obituary notice in *The Observatory*, Jan. 1913, p. 58:

The Association visited a dozen towns, and at each halt its President addressed an audience partly new, and partly composed of people who had been travelling with him for many weeks.   At each place this latter section heard with admiration a treatment of his subject wholly fresh and exactly adapted to the locality.

Such duties are always trying and it should not be forgotten that tact was necessary in a country which only two years before was still in the throes of war.

In the autumn he received the honour of being made a K.C.B.   The distinction was doubly valued as being announced to him by his friend Mr Balfour, then Prime Minister.

From 1899 to 1900 he was President of the Royal Astronomical Society. One of his last Presidential acts was the presentation of the Society's Medal to his friend M. Poincaré.

He had the unusual distinction of serving twice as President of the Cambridge Philosophical Society, once in 1890—92 and again 1911—12.

In 1891 he gave the Bakerian Lecture[1] of the Royal Society, his subject being "Tidal Prediction." This annual prælection dates from 1775 and the list of lecturers is a distinguished roll of names.

In 1897 he lectured at the Lowell Institute at Boston, and this was the origin of his book on *Tides*, published in the following year. Of this Sir Joseph Larmor says[2] that "it has taken rank with the semi-popular writings of Helmholtz and Kelvin as a model of what is possible in the exposition of a scientific subject." It has passed through three English editions, and has been translated into many foreign languages.

International Associations.

During the last ten or fifteen years of his life George was much occupied with various International bodies, e.g. the International Geodetic Association, the International Association of Academies, the International Congress of Mathematicians and the Seismological Congress.

With regard to the last named it was in consequence of George's report to the Royal Society that the British Government joined the Congress. It was however with the Geodetic Association that he was principally connected.

Sir Joseph Larmor (*Nature*, December 12, 1912) gives the following account of the origin of the Association:

> The earliest of topographic surveys, the model which other national surveys adopted and improved upon, was the Ordnance Survey of the United Kingdom. But the great trigonometrical survey of India, started nearly a century ago, and steadily carried on since that time by officers of the Royal Engineers, is still the most important contribution to the science of the figure of the earth, though the vast geodetic operations in the United States are now following it closely. The gravitational and other complexities incident on surveying among the great mountain masses of the Himalayas early demanded the highest mathematical assistance. The problems originally attacked in India by Archdeacon Pratt were afterwards virtually taken over by the Royal Society, and its secretary, Sir George Stokes, of Cambridge, became from 1864 onwards the adviser and referee of the survey as regards its scientific enterprises. On the retirement of Sir George Stokes, this position fell very largely to Sir George Darwin, whose relations with the India Office on this and other affairs remained close, and very highly appreciated, throughout the rest of his life.

> The results of the Indian survey have been of the highest importance for the general science of geodesy....It came to be felt that closer cooperation between different countries was essential to practical progress and to coordination of the work of overlapping surveys.

[1] See Prof. Brown's Memoir, p. xlix.
[2] *Nature*, 1912. See also Prof. Brown's Memoir, p. l.

The further history of George's connection with the Association is told in the words of its Secretary, Dr van d. Sande Bakhuyzen, to whom I am greatly indebted.

On the proposal of the Royal Society, the British Government, after having consulted the Director of the Ordnance Survey, in 1898, resolved upon the adhesion of Great Britain to the International Geodetic Association, and appointed as its delegate, G. H. Darwin. By his former researches and by his high scientific character, he, more than any other, was entitled to this position, which would afford him an excellent opportunity of furthering, by his recommendations, the study of theoretical geodesy.

The meeting at Stuttgart in 1898 was the first which he attended, and at that and the following conferences, Paris 1900, Copenhagen 1903, Budapest 1906, London-Cambridge 1909, he presented reports on the geodetic work in the British Empire. To Sir David Gill's report on the geodetic work in South Africa, which he delivered at Budapest, Darwin added an appendix in which he relates that the British South Africa Company, which had met all the heavy expense of the part of the survey along the 30th meridian through Rhodesia, found it necessary to make various economies, so that it was probably necessary to suspend the survey for a time. This interruption would be most unfortunate for the operations relating to the great triangulation from the Southern part of Cape Colony to Egypt, but, happily, by the cooperation of different authorities, all obstacles had been overcome and the necessary money found, so that the triangulation could be continued. So much for Sir George Darwin's communication; it is correct but incomplete, as it does not mention that it was principally by Darwin's exertions and by his personal offer of financial help that the question was solved and the continuation of this great enterprise secured.

To the different researches which enter into the scope of the Geodetic Association belong the researches on the tides, and it is natural that Darwin should be chosen as general reporter on that subject; two elaborate reports were presented by him at the conferences of Copenhagen and London.

In Copenhagen he was a member of the financial committee, and at the request of this body he presented a report on the proposal to determine gravity at sea, in which he strongly recommended charging Dr Hecker with that determination using the method of Prof. Mohn (boiling temperature of water and barometer readings). At the meeting of 1906 an interesting report was read by him on a question raised by the Geological Congress: the cooperation of the Geodetic Association in geological researches by means of the anomalies in the intensity of gravitation.

By these reports and recommendations Darwin exercised a useful influence on the activity of the Association, but his influence was to be still increased. In 1907 the Vice-president of the Association, General Zacharias, died, and the permanent committee, whose duty it was to nominate his provisional successor, chose unanimously Sir George Darwin, and this choice was confirmed by the next General Conference in London.

We cannot relate in detail his valuable cooperation as a member of the council in the various transactions of the Association, for instance on the junction of the Russian and Indian triangulations through Pamir, but we must gratefully remember his great service to the Association when, at his invitation, the delegates met in 1909 for the 16th General Conference in London and Cambridge.

With the utmost care he prepared everything to render the Conference as interesting and agreeable as possible, and he fully succeeded. Through his courtesy the foreign delegates had the opportunity of making the personal acquaintance of several members of the Geodetic staff of England and its colonies, and of other scientific men, who were invited to take part in the conference; and when after four meetings in London the delegates went to Cambridge to continue their work, they enjoyed the most cordial hospitality from Sir George and Lady Darwin, who, with her husband, procured them in Newnham Grange happy leisure hours between their scientific labours.

At this conference Darwin delivered various reports, and at the discussion on Hecker's determination of the variation of the vertical by the attraction of the moon and sun, he gave an interesting account of the researches on the same subject made by him and his brother Horace more than 20 years ago, which unfortunately failed from the bad conditions of the places of observation.

In 1912 Sir George, though already over-fatigued by the preparations for the mathematical congress in Cambridge, and the exertions entailed by it, nevertheless prepared the different reports on the geodetic work in the British Empire, but alas his illness prevented him from assisting at the conference at Hamburg, where they were presented by other British delegates. The conference thanked him and sent him its best wishes, but at the end of the year the Association had to deplore the loss of the man who in theoretical geodesy as well as in other branches of mathematics and astronomy stood in the first rank, and who for his noble character was respected and beloved by all his colleagues in the International Geodetic Association.

Sir Joseph Larmor writes[1]:

Sir George Darwin's last public appearance was as president of the fifth International Congress of Mathematicians, which met at Cambridge on August 22—28, 1912. The time for England to receive the congress having obviously arrived, a movement was initiated at Cambridge, with the concurrence of Oxford mathematicians, to send an invitation to the fourth congress held at Rome in 1908. The proposal was cordially accepted, and Sir George Darwin, as *doyen* of the mathematical school at Cambridge, became chairman of the organising committee, and was subsequently elected by the congress to be their president. Though obviously unwell during part of the meeting, he managed to discharge the delicate duties of the chair with conspicuous success, and guided with great *verve* the deliberations of the final assembly of what turned out to be a most successful meeting of that important body.

[1] *Nature*, Dec. 12, 1912.

Personal Characteristics.

His daughter, Madame Raverat, writes:

I think most people might not realise that the sense of adventure and romance was the most important thing in my father's life, except his love of work. He thought about all life romantically and his own life in particular; one could feel it in the quality of everything he said about himself. Everything in the world was interesting and wonderful to him and he had the power of making other people feel it.

He had a passion for going everywhere and seeing everything; learning every language, knowing the technicalities of every trade; and all this emphatically *not* from the scientific or collector's point of view, but from a deep sense of the romance and interest of everything. It was splendid to travel with him; he always learned as much as possible of the language, and talked to everyone; we had to see simply everything there was to be seen, and it was all interesting like an adventure. For instance at Vienna I remember being taken to a most improper music hall; and at Schönbrunn hearing from an old forester the whole secret history of the old Emperor's son. My father would tell us the stories of the places we went to with an incomparable conviction, and sense of the reality and dramaticness of the events. It is absurd of course, but in that respect he always seemed to me a little like Sir Walter Scott[1].

The books he used to read to us when we were quite small, and which we adored, were Percy's *Reliques* and the *Prologue to the Canterbury Tales*. He used often to read Shakespeare to himself, I think generally the historical plays, Chaucer, *Don Quixote* in Spanish, and all kind of books like Joinville's *Life of St Louis* in the old French.

I remember the story of the death of Gordon told so that we all cried, I think; and Gladstone could hardly be mentioned in consequence. All kinds of wars and battles interested him, and I think he liked archery more because it was romantic than because it was a game.

During his last illness his interest in the Balkan war never failed. Three weeks before his death he was so ill that the doctor thought him dying. Suddenly he rallied from the half-unconscious state in which he had been lying for many hours and the first words he spoke on opening his eyes were: "Have they got to Constantinople yet?" This was very characteristic. I often wish he was alive now, because his understanding and appreciation of the glory and tragedy of this war would be like no one else's.

His daughter Margaret Darwin writes:

He was absolutely unselfconscious and it never seemed to occur to him to wonder what impression he was making on others. I think it was this simplicity which made him so good with children. He seemed to understand their point of view and to enjoy *with* them in a way that

---

[1] Compare Mr Chesterton's *Twelve Types*, 1903, p. 190. He speaks of Scott's critic in the *Edinburgh Review*: "The only thing to be said about that critic is that he had never been a little boy. He foolishly imagined that Scott valued the plume and dagger of Marmion for Marmion's sake. Not being himself romantic, he could not understand that Scott valued the plume because it was a plume and the dagger because it was a dagger."

is not common with grown-up people.   I shall never forget how when our dog had to be killed he seemed to feel the horror of it just as I did, and how this sense of his really sharing my grief made him able to comfort me as nobody else could.

He took a transparent pleasure in the honours that came to him, especially in his membership of foreign Academies, in which he and Sir David Gill had a friendly rivalry or "race," as they called it.   I think this simplicity was one of his chief characteristics, though most important of all was the great warmth and width of his affections.   He would take endless trouble about his friends, especially in going to see them if they were lonely or ill; and he was absolutely faithful and generous in his love.

After his mother came to live in Cambridge, I believe he hardly ever missed a day in going to see her even though he might only be able to stay a few minutes.   She lived at some distance off and he was often both busy and tired.   This constancy was very characteristic.   It was shown once more in his many visits to Jim Harradine, the marker at the tennis court, on what proved to be his death-bed.

His energy and his kindness of heart were shown in many cases of distress. For instance, a guard on the Great Northern Railway was robbed of his savings by an absconding solicitor, and George succeeded in collecting some £300 for him.   In later years, when his friend the guard became bedridden, George often went to see him.   Another man whom he befriended was a one-legged man at Balsham whom he happened to notice in bicycling past.   He took the trouble to see the village authorities and succeeded in sending the man to London to be fitted with an artificial leg.

In these and similar cases there was always the touch of personal sympathy.   For instance he pensioned the widow of his gardener, and he often made the payment of her weekly allowance the excuse for a visit.

In another sort of charity he was equally kind-hearted, viz. in answering the people who wrote foolish letters to him on scientific subjects—and here as in many points he resembled his father.

His sister, Mrs Litchfield, has truly said[1] of George that he inherited his father's power of work and much of his "cordiality and warmth of nature with a characteristic power of helping others."   He resembled his father in another quality, that of modesty.   His friend and pupil E. W. Brown writes:

He was always modest about the importance of his researches. He would often wonder whether the results were worth the labour they had cost him and whether he would have been better employed in some other way.

His nephew Bernard, speaking of George's way of taking pains to be friendly and forthcoming to anyone with whom he came in contact, says:

[1] *Emma Darwin, A Century of Family Letters*, 1915, vol. II. p. 146.

He was ready to take other people's pleasantness and politeness at its apparent value and not to discount it. If they seemed glad to see him, he believed that they *were* glad. If he liked somebody, he believed that the somebody liked him, and did not worry himself by wondering whether they really did like him.

Of his energy we have evidence in the *amount* of work contained in these volumes. There was nothing dilatory about him, and here he again resembled his father who had markedly the power of doing things at the right moment, and thus avoiding waste of time and discomfort to others. George had none of a characteristic which was defined in the case of Henry Bradshaw, as "always doing something else." After an interruption he could instantly reabsorb himself in his work, so that his study was not kept as a place sacred to peace and quiet.

His wife is my authority for saying that although he got so much done, it was not by working long hours. Moreover the days that he was away from home made large gaps in his opportunities for steady application. His diaries show in another way that his researches by no means took all his time. He made a note of the books he read and these make a considerable record. Although he read much good literature with honest enjoyment, he had not a delicate or subtle literary judgment. Nor did he care for music. He was interested in travels, history, and biography, and as he could remember what he read or heard, his knowledge was wide in many directions. His linguistic power was characteristic. He read many European languages. I remember his translating a long Swedish paper for my father. And he took pleasure in the Platt Deutsch stories of Fritz Reuter.

The discomfort from which he suffered during the meeting at Cambridge of the International Congress of Mathematicians in August 1912, was in fact the beginning of his last illness. An exploratory operation showed that he was suffering from malignant disease. Happily he was spared the pain that gives its terror to this malady. His nature was, as we have seen, simple and direct with a pleasant residue of the innocence and eagerness of childhood. In the manner of his death these qualities were ennobled by an admirable and most unselfish courage. As his vitality ebbed away his affection only showed the stronger. He wished to live, and he felt that his power of work and his enjoyment of life were as strong as ever, but his resignation to the sudden end was complete and beautiful. He died on Dec. 7, 1912, and was buried at Trumpington.

HONOURS, MEDALS, DEGREES, SOCIETIES, ETC.

Order. K.C.B. 1905.

Medals[1].

   1883.   Telford Medal of the Institution of Civil Engineers.
   1884.   Royal Medal[2].
   1892.   Royal Astronomical Society's Medal.
   1911.   Copley Medal of the Royal Society.
   1912.   Royal Geographical Society's Medal.

Offices.

Fellow of Trinity College, Cambridge, and Plumian Professor in the University.

Vice-President of the International Geodetic Association, Lowell Lecturer at Boston U.S. (1897).

Member of the Meteorological and Solar Physics Committees.

Past President of the Cambridge Philosophical Society[3], Royal Astronomical Society, British Association.

Doctorates, etc. of Universities.

Oxford, Dublin, Glasgow, Pennsylvania, Padua (Socio onorario), Göttingen, Christiania, Cape of Good Hope, Moscow (honorary member).

Foreign or Honorary Membership of Academies, etc.

Amsterdam (Netherlands Academy), Boston (American Academy), Brussels (Royal Society), Calcutta (Math. Soc.), Dublin (Royal Irish Academy), Edinburgh (Royal Society), Halle (K. Leop.-Carol. Acad.), Kharkov (Math. Soc.), Mexico (Soc. "Antonio Alzate"), Moscow (Imperial Society of the Friends of Science), New York, Padua, Philadelphia (Philosophical Society), Rome (Lincei), Stockholm (Swedish Academy), Toronto (Physical Society), Washington (National Academy), Wellington (New Zealand Inst.).

Correspondent of Academies, etc. at

Acireale (Zelanti), Berlin (Prussian Academy), Buda Pest (Hungarian Academy), Frankfort (Senckenberg. Natur. Gesell.), Göttingen (Royal Society), Paris, St Petersburg, Turin, Istuto Veneto, Vienna[4].

[1] Sir George's medals are deposited in the Library of Trinity College, Cambridge.

[2] Given by the Sovereign on the nomination of the Royal Society.

[3] Re-elected in 1912.

[4] The above list is principally taken from that compiled by Sir George for the Year-Book of the Royal Society, 1912, and may not be quite complete.

It should be added that he especially valued the honour conferred on him in the publication of his collected papers by the Syndics of the University Press.

THE SCIENTIFIC WORK OF SIR GEORGE DARWIN

BY

PROFESSOR E. W. BROWN

The scientific work of Darwin possesses two characteristics which cannot fail to strike the reader who glances over the titles of the eighty odd papers which are gathered together in the four volumes which contain most of his publications. The first of these characteristics is the homogeneous nature of his investigations. After some early brief notes, on a variety of subjects, he seems to have set himself definitely to the task of applying the tests of mathematics to theories of cosmogony, and to have only departed from it when pressed to undertake the solution of practical problems for which there was an immediate need. His various papers on viscous spheroids concluding with the effects of tidal friction, the series on rotating masses of fluids, even those on periodic orbits, all have the idea, generally in the foreground, of developing the consequences of old and new assumptions concerning the past history of planetary and satellite systems. That he achieved so much, in spite of indifferent health which did not permit long hours of work at his desk, must have been largely due to this single aim.

The second characteristic is the absence of investigations undertaken for their mathematical interest alone; he was an applied mathematician in the strict and older sense of the word. In the last few decades another school of applied mathematicians, founded mainly by Poincaré, has arisen, but it differs essentially from the older school. Its votaries have less interest in the phenomena than in the mathematical processes which are used by the student of the phenomena. They do not expect to examine or predict physical events but rather to take up the special classes of functions, differential equations or series which have been used by astronomers or physicists, to examine their properties, the validity of the arguments and the limitations which must be placed on the results. Occasionally theorems of great physical importance will emerge, but from the primary point of view of the investigations these are subsidiary results. Darwin belonged essentially to the school which studies the phenomena by the most convenient mathematical methods. Strict logic in the modern sense is not applied nor is it necessary, being replaced in most cases by intuition which guides the investigator through the dangerous places. That the new school has done great service to both pure and applied mathematics can hardly be doubted, but the two points of view of the subject will but rarely be united in the same man if much progress in either direction is to be made. Hence we do not find and do not expect to find in Darwin's work developments from the newer point of view.

At the same time, he never seems to have been affected by the problem-solving habits which were prevalent in Cambridge during his undergraduate days and for some time later. There was then a large number of mathematicians brought up in the Cambridge school whose chief delight was the discovery of a problem which admitted of a neat mathematical solution. The chief leaders were, of course, never very seriously affected by this attitude; they had larger objects in view, but the temptation to work out a problem, even one of little physical importance, when it would yield to known mathematical processes, was always present. Darwin kept his aim fixed. If the problem would not yield to algebra he has recourse to arithmetic; in either case he never seemed to hesitate to embark on the most complicated computations if he saw a chance of attaining his end. The papers on ellipsoidal harmonic analysis and periodic orbits are instructive examples of the labour which he would undertake to obtain a knowledge of physical phenomena.

One cannot read any of his papers without also seeing another feature, his preference for quantitative rather than qualitative results. If he saw any possibility of obtaining a numerical estimate, even in his most speculative work, he always made the necessary calculations. His conclusions thus have sometimes an appearance of greater precision than is warranted by the degree of accuracy of the data. But Darwin himself was never misled by his numerical conclusions, and he is always careful to warn his readers against laying too great a stress on the numbers he obtains.

In devising processes to solve his problems, Darwin generally adopted those which would lead in a straightforward manner to the end he had in view. Few "short cuts" are to be found in his memoirs. He seems to have felt that the longer processes often brought out details and points of view which would otherwise have been concealed or neglected. This is particularly evident in the papers on Periodic Orbits. In the absence of general methods for the discovery and location of the curves, his arithmetic showed classes of orbits which would have been difficult to find by analysis, and it had a further advantage in indicating clearly the various changes which the members of any class undergo when the parameter varies. Yet, in spite of the large amount of numerical work which is involved in many of his papers, he never seemed to have any special liking for either algebraic or numerical computation; it was something which "had to be done." Unlike J. C. Adams and G. W. Hill, who would often carry their results to a large number of places of decimals, Darwin would find out how high a degree of accuracy was necessary and limit himself to it.

The influence which Darwin exerted has been felt in many directions. The exhibition of the necessity for quantitative and thorough analysis of the problems of cosmogony and celestial mechanics has been perhaps one of his chief contributions. It has extended far beyond the work of the pupils who were directly inspired by him. While speculations and the framing of new hypotheses must continue, but little weight is now attached to those which are defended by general reasoning alone. Conviction fails, possibly because it is recognised that the human mind cannot reason accurately in these questions without the aids furnished by mathematical symbols, and in any case language often fails to carry fully the argument of the writer as against the exact implications of mathematics. If for no other reason, Darwin's work marks an epoch in this respect.

To the pupils who owed their first inspiration to him, he was a constant friend. First meeting them at his courses on some geophysical or astronomical subject, he soon dropped the formality of the lecture-room, and they found themselves before long going to see him continually in the study at Newnham Grange. Who amongst those who knew him will fail to remember the sight of him seated in an armchair with a writing board and papers strewn about the table and floor, while through the window were seen glimpses of the garden filled in summer time with flowers? While his lectures in the class-room were always interesting and suggestive, the chief incentive, at least to the writer who is proud to have been numbered amongst his pupils and friends, was conveyed through his personality. To have spent an hour or two with him, whether in discussion on "shop" or in general conversation, was always a lasting inspiration. And the personal attachment of his friends was strong; the gap caused by his death was felt to be far more than a loss to scientific progress. Not only the solid achievements contained in his published papers, but the spirit of his work and the example of his life will live as an enduring memorial of him.

*   *   *   *   *   *

Darwin's first five papers, all published in 1875, are of some interest as showing the mechanical turn of his mind and the desire, which he never lost, for concrete illustrations of whatever problem might be interesting him. A Peaucellier's cell is shown to be of use for changing a constant force into one varying inversely as the square of the distance, and it is applied to the description of equipotential lines. A method for describing graphically the second elliptic integral and one for map projection on the face of a polyhedron are also given. There are also a few other short papers of the same kind but of no special importance, and Darwin says that he only included them in his collected works for the sake of completeness.

His first important contributions obviously arose through the study of the works of his predecessors, and though of the nature of corrections to previously accepted or erroneous ideas, they form definite additions to the subject of cosmogony. The opening paragraph of the memoir "On the influence of geological changes in the earth's axis of rotation" describes the situation which prompted the work. "The subject of the fixity or mobility of the earth's axis of rotation in that body, and the possibility of variations in the obliquity of the ecliptic, have from time to time attracted the notice of mathematicians and geologists. The latter look anxiously for some grand cause capable of producing such an enormous effect as the glacial period. Impressed by the magnitude of the phenomenon, several geologists have postulated a change of many degrees in the obliquity of the ecliptic and a wide variability in the position of the poles on the earth; and this, again, they have sought to refer back to the upheaval and subsidence of continents." He therefore subjects the hypothesis to mathematical examination under various assumptions which have either been put forward by geologists or which he considers à priori probable. The conclusion, now well known to astronomers, but frequently forgotten by geologists even at the present time, is against any extensive wanderings of the pole during geological times. "Geologists and biologists," writes Professor Barrell[1], "may array facts which suggest such hypotheses, but the testing of their possibility is really a problem of mathematics, as much as are the movements of precession, and orbital perturbations. Notwithstanding this, a number of hypotheses concerning polar migration have been ingeniously elaborated and widely promulgated without their authors submitting them to these final tests, or in most cases even perceiving that an accordance with the known laws of mechanics was necessary....A reexamination of these assumptions in the light of forty added years of geological progress suggests that the actual changes have been much less and more likely to be limited to a fraction of the maximum limits set by Darwin. His paper seems to have checked further speculation upon this subject in England, but, apparently unaware of its strictures, a number of continental geologists and biologists have carried forward these ideas of polar wandering to the present day. The hypotheses have grown, each creator selecting facts and building up from his particular assortment a fanciful hypothesis of polar migration unrestrained even by the devious paths worked out by others." The methods used by Darwin are familiar to those who investigate problems connected with the figure of the earth, but the whole paper is characteristic of his style in the careful arrangement of the assumptions, the conclusions deduced therefrom, the frequent reduction to numbers and the summary giving the main results.

It is otherwise interesting because it was the means of bringing Darwin into close connection with Lord Kelvin, then Sir William Thomson. The

---

[1] *Science*, Sept. 4, 1914, p. 333.

latter was one of the referees appointed by the Royal Society to report on it, and, as Darwin says, " He seemed to find that on these occasions the quickest way of coming to a decision was to talk over the subject with the author himself—at least this was frequently so as regards myself." Through his whole life Darwin, like many others, prized highly this association, and he considered that his whole work on cosmogony "may be regarded as the scientific outcome of our conversation of the year 1877 ; but," he adds, "for me at least science in this case takes the second place."

Darwin at this time was thirty-two years old. In the three years since he started publication fourteen memoirs and short notes, besides two statistical papers on marriage between first cousins, form the evidence of his activity. He seems to have reached maturity in his mathematical power and insight into the problems which he attacked without the apprenticeship which is necessary for most investigators. Probably the comparatively late age at which he began to show his capacity in print may have something to do with this. Henceforth development is rather in the direction of the full working out of his ideas than growth of his powers. It seems better therefore to describe his further scientific work in the manner in which he arranged it himself, by subject instead of in chronological order. And here we have the great advantage of his own comments, made towards the end of his life when he scarcely hoped to undertake any new large piece of work. Frequent quotation will be made from these remarks which occur in the prefaces to the volumes, in footnotes and in his occasional addresses.

The following account of the Earth-Moon series of papers is taken bodily from the Notice in the *Proceedings of the Royal Society* [1] by Mr S. S. Hough, who was himself one of Darwin's pupils.

" The conclusions arrived at in the paper referred to above were based on the assumption that throughout geological history, apart from slow geological changes, the Earth would rotate sensibly as if it were rigid. It is shown that a departure from this hypothesis might possibly account for considerable excursions of the axis of rotation within the Earth itself, though these would be improbable, unless, indeed, geologists were prepared to abandon the view 'that where the continents now stand they have always stood'; but no such effect is possible with respect to the direction of the Earth's axis in space. Thus the present condition of obliquity of the Earth's equator could in no way be accounted for as a result of geological change, and a further cause had to be sought. Darwin foresaw a possibility of obtaining an explanation in the frictional resistance to which the tidal oscillations of the mobile parts of a planet must be subject. The investigation of this hypothesis gave rise to a remarkable series of papers of far-reaching consequence in theories of cosmogony and of the present constitution of the Earth.

[1] Vol. 89 A, p. i.

"In the first of these papers, which is of preparatory character, 'On the Bodily Tides of Viscous and Semi-elastic Spheroids, and on the Ocean Tides on a Yielding Nucleus' (*Phil. Trans.*, 1879, vol. 170), he adapts the analysis of Sir William Thomson, relating to the tidal deformations of an elastic sphere, to the case of a sphere composed of a viscous liquid or, more generally, of a material which partakes of the character either of a solid or a fluid according to the nature of the strain to which it is subjected. For momentary deformations it is assumed to be elastic in character, but the elasticity is considered as breaking down with continuation of the strain in such a manner that under very slow variations of the deforming forces it will behave sensibly as if it were a viscous liquid. The exact law assumed by Darwin was dictated rather by mathematical exigencies than by any experimental justification, but the evidence afforded by the flow of rocks under continuous stress indicates that it represents, at least in a rough manner, the mechanical properties which characterise the solid parts of the Earth.

"The chief practical result of this paper is summed up by Darwin himself by saying that it is strongly confirmatory of the view already maintained by Kelvin that the existence of ocean tides, which would otherwise be largely masked by the yielding of the ocean bed to tidal deformation, points to a high effective rigidity of the Earth as a whole. Its value, however, lies further in the mathematical expressions derived for the reduction in amplitude and retardation in phase of the tides resulting from viscosity which form the starting-point for the further investigations to which the author proceeded.

"The retardation in phase or 'lag' of the tide due to the viscosity implies that a spheroid as tidally distorted will no longer present a symmetrical aspect as if no such cause were operative. The attractive forces on the nearer and more distant parts will consequently form a non-equi-librating system with resultant couples tending to modify the state of rotation of the spheroid about its centre of gravity. The action of these couples, though exceedingly small, will be cumulative with lapse of time, and it is their cumulative effects over long intervals which form the subject of the next paper, 'On the Precession of a Viscous Spheroid and on the Remote History of the Earth' (*Phil. Trans.*, 1879, vol. 170, Part II, pp. 447—530). The case of a single disturbing body (the Moon) is first considered, but it is shown that if there are two such bodies raising tidal disturbances (the Sun and Moon) the conditions will be materially modified from the superposed results of the two disturbances considered separately. Under certain conditions of viscosity and obliquity the obliquity of the ecliptic will increase, and under others it will diminish, but the analysis further yields 'some remarkable results as to the dynamical stability or instability of the system...for moderate degrees of viscosity, the position of zero obliquity is unstable, but there is a position of stability at a high obliquity. For large viscosities the position of zero obliquity becomes stable, and (except for a very close approximation to rigidity) there is an unstable position at a larger obliquity, and again a stable one at a still larger one.'

"The reactions of the tidal disturbing force on the motion of the Moon are next considered, and a relation derived connecting that portion of the apparent secular acceleration of the Moon's mean motion, which cannot be otherwise accounted for by theory, with the heights and retardations of the several bodily tides in the Earth. Various hypotheses are discussed, but with the conclusion that insufficient evidence is available to form 'any estimate having any pretension to accuracy...as to the present rate of change due to tidal friction.'

"But though the time scale involved must remain uncertain, the nature of the physical changes that are taking place at the present time is practically free from obscurity. These involve a gradual increase in the length of the day, of the month, and of the obliquity of the ecliptic, with a gradual recession of the Moon from the Earth. The most striking result is that these changes can be traced backwards in time until a state is reached when the Moon's centre would be at a distance of only about 6000 miles from the Earth's surface, while the day and month would be of equal duration, estimated at 5 hours 36 minutes. The minimum time which can have elapsed since this condition obtained is further estimated at about 54 million years. This leads to the inevitable conclusion that the Moon and Earth at one time formed parts of a common mass and raises the question of how and why the planet broke up. The most probable hypothesis appeared to be that, in accordance with Laplace's nebular hypothesis, the planet, being partly or wholly fluid, contracted, and thus rotated faster and faster, until the ellipticity became so great that the equilibrium was unstable.

"The tentative theory put forward by Darwin, however, differs from the nebular hypothesis of Laplace in the suggestion that instability might set in by the rupture of the body into two parts rather than by casting off a ring of matter, somewhat analogous to the rings of Saturn, to be afterwards consolidated into the form of a satellite.

"The mathematical investigation of this hypothesis forms a subject to which Darwin frequently reverted later, but for the time he devoted himself to following up more minutely the motions which would ensue after the supposed planet, which originally consisted of the existing Earth and Moon in combination, had become detached into two separate masses. In the final section of a paper 'On the Secular Changes in the Elements of the Orbit of a Satellite revolving about a Tidally Distorted Planet' (*Phil. Trans.*, 1880, vol. 171), Darwin summarises the results derived in his different memoirs. Various factors ignored in the earlier investigations, such as the eccentricity and inclination of the lunar orbit, the distribution of the heat generated by tidal friction and the effects of inertia, were duly considered and a complete history traced of the evolution resulting from tidal friction of a system originating as two detached masses nearly in contact with one another and rotating nearly as though they were parts of one rigid body. Starting with the numerical data suggested by the Earth-Moon System, 'it is only necessary to postulate a sufficient lapse of time, and that there is not enough matter diffused through space to resist materially the motions of the Moon and Earth,' when 'a system would necessarily be developed which would bear a strong resemblance to our own.' 'A theory, reposing on *verae causae*, which brings into quantitative correlation the lengths of the present day and month, the obliquity of the ecliptic, and the inclination and eccentricity of the lunar orbit, must, I think, have strong claims to acceptance.'

"Confirmation of the theory is sought and found, in part at least, in the case of other members of the Solar System which are found to represent various stages in the process of evolution indicated by the analysis.

"The application of the theory of tidal friction to the evolution of the Solar System and of planetary sub-systems other than the Earth-Moon System is, however, reconsidered later, 'On the Tidal Friction of a Planet attended by Several Satellites, and on the Evolution of the Solar System' (*Phil. Trans.*, 1882, vol. 172). The conclusions drawn in this paper are that the Earth-Moon System forms a unique example within the Solar System of its particular mode of evolution. While tidal friction may perhaps be invoked to throw light on the distribution of the satellites among the several planets, it is very improbable that it has figured as the dominant cause of change of the other planetary systems or in the Solar System itself."

For some years after this series of papers Darwin was busy with practical tidal problems but he returned later "to the problems arising in connection with the genesis of the Moon, in accordance with the indications previously arrived at from the theory of tidal friction. It appeared to be of interest to trace back the changes which would result in the figures of the Earth and Moon, owing to their mutual attraction, as they approached one another. The analysis is confined to the consideration of two bodies supposed constituted of homogeneous liquid. At considerable distances the solution of the problem thus presented is that of the equilibrium theory of the tides, but, as the masses are brought nearer and nearer together, the approximations available for the latter problem cease to be sufficient. Here, as elsewhere, when the methods of analysis could no longer yield algebraic results, Darwin boldly proceeds to replace his symbols by numerical quantities, and thereby succeeds in tracing, with considerable approximation, the forms which such figures would assume when the two masses are nearly in contact. He even carries the investigation farther, to a stage when the two masses in part overlap. The forms obtained in this case can only be regarded as satisfying the analytical, and not the true physical conditions of the problem, as, of course, two different portions of matter cannot occupy the same space. They, however, suggest that, by a very slight modification of conditions, a new form could be found, which would fulfil all the conditions, in which the two detached masses are united into a single mass, whose shape has been variously described as resembling that of an hour-glass, a dumb-bell, or a pear. This confirms the suggestion previously made that the origin of the Moon was to be sought in the rupture of the parent planet into two parts, but the theory was destined to receive a still more striking confirmation from another source.

"While Darwin was still at work on the subject, there appeared the great memoir by M. Poincaré, 'Sur l'équilibre d'une masse fluide animée d'un mouvement de rotation' (*Acta Math.*, vol. 7).

"The figures of equilibrium known as Maclaurin's spheroid and Jacobi's ellipsoid were already familiar to mathematicians, though the conditions of stability, at least of the latter form, were not established. By means of analysis of a masterly character, Poincaré succeeded in enunciating and applying to this problem the principle of exchange of stabilities. This principle may be briefly indicated as follows: Imagine a dynamical system such as a rotating liquid planet to be undergoing evolutionary change such as would result from a gradual condensation of its mass through cooling. Whatever be the varying element to which the evolutionary changes may be referred, it may be possible to define certain relatively simple modes of motion, the features associated with which will, however, undergo continuous evolution. If the existence of such modes has been established, M. Poincaré shows that the investigation of their persistence or 'stability' may be made to depend on the evaluation of certain related quantities which he defines as coefficients of stability. The latter quantities will be subject to evolutionary change, and it may happen that in the course of such change one or more of them assumes a zero value. Poincaré shows that such an occurrence indicates that the particular mode of motion under consideration coalesces at this stage with one other mode which likewise has a vanishing coefficient of stability. Either mode will, as a rule, be possible before the change, but whereas one will be stable the other will be unstable. The same will be true after the change, but there will be an interchange of stabilities, whereby that which was previously stable will become unstable, and *vice versâ*. An illustration of this principle was found in the case of the spheroids of Maclaurin and the ellipsoids of Jacobi. The former in the earlier stages of evolution will represent a stable condition, but as the ellipticity of surface increases a stage is reached where it ceases to be stable and becomes unstable.

At this stage it is found to coalesce with Jacobi's form which involves in its further development an ellipsoid with three unequal axes. Poincaré shows that the latter form possesses in its earlier stages the requisite elements of stability, but that these in their turn disappear in the later developments. In accordance with the principle of exchange of stabilities laid down by him, the loss of stability will occur at a stage where there is coalescence with another form of figure, to which the stability will be transferred, and this form he shows at its origin resembles the pear which had already been indicated by Darwin's investigation. The supposed pear-shaped figure was thus arrived at by two entirely different methods of research, that of Poincaré tracing the processes of evolution forwards and that of Darwin proceeding backwards in time.

"The chain of evidence was all but complete; it remained, however, to consider whether the pear-shaped figure indicated by Poincaré, stable in its earlier forms, could retain its stability throughout the sequence of changes necessary to fill the gap between these forms and the forms found by Darwin.

"In later years Darwin devoted much time to the consideration of this problem. Undeterred by the formidable analysis which had to be faced, he proceeded to adapt the intricate theory of Ellipsoidal Harmonics to a form in which it would admit of numerical application, and his paper 'Ellipsoid Harmonic Analysis' (*Phil. Trans.*, A, 1901, vol. 197), apart from the application for which it was designed, in itself forms a valuable contribution to this particular branch of analysis. With the aid of these preliminary investigations he succeeded in tracing with greater accuracy the form of the pear-shaped figure as established by Poincaré, 'On the Pear-shaped Figure of Equilibrium of a Rotating Mass of Liquid' (*Phil. Trans.*, A, 1901, vol. 198), and, as he considered, in establishing its stability, at least in its earlier forms. Some doubt, however, is expressed as to the conclusiveness of the argument employed, as simultaneous investigations by M. Liapounoff pointed to an opposite conclusion. Darwin again reverts to this point in a further paper 'On the Figure and Stability of a Liquid Satellite' (*Phil. Trans.*, A, 1906, vol. 206), in which is considered the stability of two isolated liquid masses in the stage at which they are in close proximity, i.e., the condition which would obtain, in the Earth-Moon System, shortly after the Moon had been severed from the Earth. The ellipsoidal harmonic analysis previously developed is then applied to the determination of the approximately ellipsoidal forms which had been indicated by Roche. The conclusions arrived at seem to point, though not conclusively, to instability at the stage of incipient rupture, but in contradistinction to this are quoted the results obtained by Jeans, who considered the analogous problems of the equilibrium and rotation of infinite rotating cylinders of liquid. This problem is the two-dimensional analogue of the problems considered by Darwin and Poincaré, but involves far greater simplicity of the conditions. Jeans finds solutions of his problem strictly analogous to the spheroids of Maclaurin, the ellipsoids of Jacobi, and the pear of Poincaré, and is able to follow the development of the latter until the neck joining the two parts has become quite thin. He is able to establish conclusively that the pear is stable in its early stages, while there is no evidence of any break in the stability up to the stage when it divides itself into two parts."

Darwin's own final comments on this work next find a place here. He is writing the preface to the second volume of his Collected Works in 1908, after which time nothing new on the subject came from his pen. "The observations of Dr Hecker," he says, "and of others do not afford evidence of any considerable amount of retardation in the tidal oscillations of the solid earth, for, within the limits of error of observation, the phase of the oscillation appears to be the same as if the earth were purely elastic. Then again modern researches in the lunar theory show that the secular acceleration of the moon's mean motion is so nearly explained by means of pure gravitation as to leave but a small residue to be referred to the effects of tidal friction. We are thus driven to believe that at present tidal friction is producing its inevitable effects with extreme slowness. But we need not therefore hold that the march of events was always so leisurely, and if the earth was ever wholly or in large part molten, it cannot have been the case.

"In any case frictional resistance, whether it be much or little and whether applicable to the solid planet or to the superincumbent ocean, is a true cause of change....

"For the astronomer who is interested in cosmogony the important point is the degree of applicability of the theory as a whole to celestial evolution. To me it seems that the theory has rather gained than lost in the esteem of men of science during the last 25 years, and I observe that several writers are disposed to accept it as an established acquisition to our knowledge of cosmogony.

"Undue weight has sometimes been laid on the exact numerical values assigned for defining the primitive configurations of the earth and moon. In so speculative a matter close accuracy is unattainable, for a different theory of frictionally retarded tides would inevitably lead to a slight difference in the conclusion; moreover such a real cause as the secular increase in the masses of the earth and moon through the accumulation of meteoric dust, and possibly other causes, are left out of consideration.

"The exact nature of the process by which the moon was detached from the earth must remain even more speculative. I suggested that the fission of the primitive planet may have been brought about by the synchronism of the solar tide with the period of the fundamental free oscillation of the planet, and the suggestion has received a degree of attention which I never anticipated. It may be that we shall never attain to a higher degree of certainty in these obscure questions than we now possess, but I would maintain that we may now hold with confidence that the moon originated by a process of fission from the primitive planet, that at first she revolved in an orbit close to the present surface of the earth, and that tidal friction has been the principal agent which transformed the system to its present configuration.

"The theory for a long time seemed to lie open to attack on the ground that it made too great demands on time, and this has always appeared to me the greatest difficulty in the way of its acceptance. If we were still compelled to assent to the justice of Lord Kelvin's views as to the period of time which has elapsed since the earth solidified, and as to the age of the solar system, we should also have to admit the theory of evolution under tidal influence as inapplicable to its full extent. Lord Kelvin's contributions to cosmogony have been of the first order of importance, but his arguments on these points no longer carry conviction with them. Lord Kelvin contended that the actual distribution of land and sea proves that the planet solidified at a time when the day had nearly its present length. If this were true the effects of tidal friction relate to a period antecedent to the solidification. But I have always felt convinced that the earth would adjust its ellipticity to its existing speed of rotation with close approximation."

After some remarks concerning the effects of the discovery of radio-activity and the energy resident in the atom on estimates of geological time, he continues, "On the whole then it may be maintained that deficiency of time does not, according to our present state of knowledge, form a bar to the full acceptability of the theory of terrestrial evolution under the influence of tidal friction.

"It is very improbable that tidal friction has been the dominant cause of change in any of the other planetary sub-systems or in the solar system itself, yet it seems to throw light on the distribution of the satellites amongst the several planets. It explains the identity of the rotation of the moon with her orbital motion, as was long ago pointed out by Kant and Laplace, and it tends to confirm the correctness of the observations according to which Venus always presents the same face to the sun."

Since this was written much information bearing on the point has been gathered from the stellar universe. The curious curves of light-changes in certain classes of spectroscopic binaries have been well explained on the assumption that the two stars are close together and under strong tidal distortion. Some of these, investigated on the same hypothesis, even seem to be in actual contact. In chap. xx of the third edition (1910) of his book on the Tides, Darwin gives a popular summary of this evidence which had

in the interval been greatly extended by the discovery and application of
the hypothesis to many other similar systems. In discussing the question
Darwin sets forth a warning. He points out that most of the densities
which result from the application of the tidal theory are very small compared
with that of the sun, and he concludes that these stars are neither homogeneous nor incompressible. Hence the figures calculated for homogeneous
liquid can only be taken to afford a general indication of the kind of figure
which we might expect to find in the stellar universe.

Perhaps Darwin's greatest service to cosmogony was the successful effort
which he made to put hypotheses to the test of actual calculation. Even
though the mathematical difficulties of the subject compel the placing of
many limitations which can scarcely exist in nature, yet the solution of even
these limited problems places the speculator on a height which he cannot
hope to attain by doubtful processes of general reasoning. If the time
devoted to the framing and setting forth of cosmogonic hypotheses by various
writers had been devoted to the accurate solution of some few problems, the
newspapers and popular scientific magazines might have been less interesting
to their readers, but we should have had more certain knowledge of our
universe. Darwin himself engaged but little in speculations which were
not based on observations or precise conclusions from definitely stated
assumptions, and then only as suggestions for further problems to be
undertaken by himself or others. And this view of progress he communicated to his pupils, one of whom, Mr J. H. Jeans, as mentioned above, is
continuing with success to solve these gravitational problems on similar
lines.

The nebular hypothesis of Kant and Laplace has long held the field as
the most probable mode of development of our solar system from a nebula.
At the present time it is difficult to say what are its chief features. Much
criticism has been directed towards every part of it, one writer changing
a detail here, another there, and still giving to it the name of the best known
exponent. The only salient point which seems to be left is the main hypothesis that the sun, planets and satellites were somehow formed during the
process of contraction of a widely diffused mass of matter to the system as
we now see it. Some writers, including Darwin himself, regard a gaseous
nebula contracting under gravitation as the essence of Laplace's hypotheses,
distinguishing this condition from that which originates in the accretion
of small masses. Others believe that both kinds of matter may be present.
After all it is only a question of a name, but it is necessary in a discussion to
know what the name means.

Darwin's paper, "The mechanical conditions of a swarm of meteorites,"
is an attempt to show that, with reasonable hypotheses, the nebula and the
small masses under contraction by collisions may have led to the same result.

In his preface to volume IV he says with respect to this paper: "Cosmogonists are of course compelled to begin their survey of the solar system at some arbitrary stage of its history, and they do not, in general, seek to explain how the solar nebula, whether gaseous or meteoritic, came to exist. My investigation starts from the meteoritic point of view, and I assume the meteorites to be moving indiscriminately in all directions. But the doubt naturally arises as to whether at any stage a purely chaotic motion of the individual meteorites could have existed, and whether the assumed initial condition ought not rather to have been an aggregate of flocks of meteorites moving about some central condensation in orbits which intersect one another at all sorts of angles. If this were so the chaos would not be one consisting of individual stones which generate a quasi-gas by their collisions, but it would be a chaos of orbits. But it is not very easy to form an exact picture of this supposed initial condition, and the problem thus seems to elude mathematical treatment. Then again have I succeeded in showing that a pair of meteorites in collision will be endowed with an effective elasticity? If it is held that the chaotic motion and the effective elasticity are quite imaginary, the theory collapses. It should however be remarked that an infinite gradation is possible between a chaos of individuals and a chaos of orbits, and it cannot be doubted that in most impacts the colliding stones would glance from one another. It seems to me possible, therefore, that my two fundamental assumptions may possess such a rough resemblance to truth as to produce some degree of similitude between the life-histories of gaseous and meteoritic nebulae. If this be so the Planetesimal Hypothesis of Chamberlain and Moulton is nearer akin to the Nebular Hypothesis than the authors of the former seem disposed to admit.

"Even if the whole of the theory could be condemned as futile, yet the paper contains an independent solution of the problem of Lane and Ritter; and besides the attempt to discuss the boundary of an atmosphere, where the collisions have become of vanishing rarity, may still perhaps be worth something."

In writing concerning the planetesimal hypothesis, Darwin seems to have forgotten that one of its central assumptions is the close approach of two stars which by violent tidal action drew off matter in spiral curves which became condensed into the attendants of each. This is, in fact, one of the most debatable parts of the hypothesis, but one on which it is possible to get evidence from the distribution of such systems in the stellar system. Controversy on the main issue is likely to exist for many years to come.

Quite early in his career Darwin was drawn into practical tidal problems by being appointed on a Committee of the British Association with Adams, to coordinate and revise previous reports drawn up by Lord Kelvin. He evidently felt that the whole subject of practical analysis of tidal observations needed to be set forth in full and made clear. His first report consequently contains a development of the equilibrium theory of the Tides, and later, after a careful analysis of each harmonic component, it proceeds to outline in detail the methods which should be adopted to obtain the constants of each component from theory or observation, as the case needed. Schedules and forms of reduction are given with examples to illustrate their use.

There are in reality two principal practical problems to be considered. The one is the case of a port with much traffic, where it is possible to obtain tide heights at frequent intervals and extending over a long period. While the accuracy needed usually corresponds to the number of observations, it is always assumed that the ordinary methods of harmonic analysis by which all other terms but that considered are practically eliminated can be applied; the corrections when this is not the case are investigated and applied. The other problem is that of a port infrequently visited, so that we have only a short series of observations from which to obtain the data for the computation of future tides. The possible accuracy here is of course lower than in the former case but may be quite sufficient when the traffic is light. In his third report Darwin takes up this question. The main difficulty is the separation of tides which have nearly the same period and which could not be disentangled by harmonic analysis of observations extending over a very few weeks. Theory must therefore be used, not only to obtain the periods, but also to give some information about the amplitudes and phases if this separation is to be effected. The magnitude of the tide-generating force is used for the purpose. Theoretically this should give correct results, but it is often vitiated by the form of the coast line and other circumstances depending on the irregular shape of the water boundary. Darwin shows however that fair prediction can generally be obtained; the amount of numerical work is of course much smaller than in the analysis of a year's observations. This report was expanded by Darwin into an article on the Tides for the *Admiralty Scientific Manual*.

Still another problem is the arrangement of the analysis when times and heights of high and low water alone are obtainable; in the previous papers the observations were supposed to be hourly or obtained from an automatically recording tide-gauge. The methods to be used in this case are of course well known from the mathematical side; the chief problem is to reduce the arithmetical work and to put the instructions into such a form that the ordinary computer may use them mechanically. The problem was worked out by Darwin in 1890, and forms the subject of a long paper in the *Proceedings of the Royal Society*.

A little later he published the description of his now well known abacus, designed to avoid the frequent rewriting of the numbers when the harmonic analysis for many different periods is needed. Much care was taken to obtain the right materials. The real objection to this, and indeed to nearly all the methods devised for the purpose, is that the arrangement and care of the mechanism takes much longer time than the actual addition of the numbers after the arrangement has been made. In this description however there are more important computing devices which reduce the time of computation to something like one-fifth of that required by the previous methods. The principal of these is the one in which it is shown how a single set of summations of 9000 hourly values can be made to give a good many terms, by dividing the sums into proper groups and suitably treating them.

Another practical problem was solved in his Bakerian Lecture "On Tidal Prediction." In a previous paper, referred to above, Darwin had shown how the tidal constants of a port might be obtained with comparatively little expense from a short series of high and low water observations. These, however, are of little value unless the port can furnish the funds necessary to predict the future times and heights of the tides. Little frequented ports can scarcely afford this, and therefore the problem of replacing such predictions by some other method is necessary for a complete solution. "The object then," says Darwin, "of the present paper, is to show how a general tide-table, applicable for all time, may be given in such a form that anyone, with an elementary knowledge of the *Nautical Almanac*, may, in a few minutes, compute two or three tides for the days on which they are required. The tables will also be such that a special tide-table for any year may be computed with comparatively little trouble."

This, with the exception of a short paper dealing with the Tides in the Antarctic as shown by observations made on the *Discovery*, concludes Darwin's published work on practical tidal problems. But he was constantly in correspondence about the subject, and devoted a good deal of time to government work and to those who wrote for information.

In connection with these investigations it was natural that he should turn aside at times to questions of more scientific interest. Of these the fortnightly tide is important because by it some estimate may be reached as to the earth's rigidity. The equilibrium theory while effective in giving the periods only for the short-period tides is much more nearly true for those of long period. Hence, by a comparison of theory and observation, it is possible to see how much the earth yields to distortion produced by the moon's attraction. Two papers deal with this question. In the first an attempt is made to evaluate the corrections to the equilibrium theory caused by the continents; this involves an approximate division of the land and sea surfaces into blocks to which calculation may be applied. In the second tidal observations from various parts of the earth are gathered together for comparison with the theoretical values. As a result, Darwin obtains the oft-quoted expression for the rigidity of the earth's mass, namely, that it is effectively about that of steel. An attempt made by George and Horace Darwin to measure the lunar disturbance of gravity by means of the pendulum is in reality another approach to the solution of the same problem. The attempt failed mainly on account of the local tremors which were produced by traffic and other causes. Nevertheless the two reports contain much that is still interesting, and their value is enhanced by a historical account of previous attempts on the same lines. Darwin had the satisfaction of knowing that this method was later successful in the hands of Dr Hecker whose results confirmed his first estimate. Since his death the remarkable experiment of Michelson[1] with a pipe partly filled with water has given a precision to the determination of this constant which much exceeds that of the older methods; he concludes that the rigidity and viscosity are at least equal to and perhaps exceed those of steel.

It is here proper to refer to Darwin's more popular expositions of the work of himself and others. He wrote several articles on Tides, notably for the *Encyclopaedia Britannica* and for the *Encyclopaedie der Mathematischen Wissenschaften*, but he will be best remembered in this connection for his volume *The Tides* which reached its third edition not long before his death. The origin of it was a course of lectures in 1897 before the Lowell Institute of Boston, Massachusetts. An attempt to explain the foundations and general developments of tidal theory is its main theme. It naturally leads on to the subject of tidal friction and the origin of the moon, and therewith are discussed numerous questions of cosmogony. From the point of view of the mathematician, it is not only clear and accurate but gives the impression, in one way, of a *tour de force*. Although Darwin rarely has to ask the reader to accept his conclusions without some description of the nature of the argument by which they are reached, there is not a single algebraic symbol in the whole volume, except in one short footnote where, on a minor detail, a little algebra is used. The achievement of this, together with a clear exposition, was no light task, and there are few examples to be found in the history of mathematics since the first and most remarkable of all, Newton's translation of the effects of gravitation into geometrical reasoning. *The Tides* has been translated into German (two editions), Hungarian, Italian and Spanish.

In 1877 the two classical memoirs of G. W. Hill on the motion of the moon were published. The first of these, *Researches in the Lunar Theory*, contains so much of a pioneer character that in writing of any later work on celestial mechanics it is impossible to dismiss it with a mere notice. One portion is directly concerned with a possible mode of development of the lunar theory and the completion of the first step in the process. The usual

[1] *Astrophysical Journal*, March, 1914.

method of procedure has been to consider the problem of three bodies as an extension of the case of two bodies in which the motion of one round the other is elliptic. Hill, following a suggestion of Euler which had been worked out by the latter in some detail, starts to treat the problem as a very special particular case of the problem of three bodies. One of them, the earth, is of finite mass; the second, the sun, is of infinite mass and at an infinite distance but is revolving round the former with a finite and constant angular velocity. The third, the moon, is of infinitesimal mass, but moves at a finite distance from the earth. Stated in this way, the problem of the moon's motion appears to bear no resemblance to reality. It is, however, nothing but a limiting case where certain constants, which are small in the case of the actual motion, have zero values. The sun is actually of very great mass compared with the earth, it is very distant as compared with the distance of the moon, its orbit round the earth (or *vice versâ*) is nearly circular, and the moon's mass is small compared with that of the earth. The differential equations which express the motion of the moon under these limitations are fairly simple and admit of many transformations.

Hill simplifies the equations still further, first by supposing the moon so started that it always remains in the same fixed plane with the earth and the sun (its actual motion outside this plane is small). He then uses moving rectangular axes one of which always points in the direction of the sun. Even with all these limitations, the differential equations possess many classes of solutions, for there will be four arbitrary constants in the most general values of the coordinates which are to be derived in the form of a doubly infinite series of harmonic terms. His final simplification is the choice of one of these classes obtained by giving a zero value to one of the arbitrary constants; in the moon's motion this constant is small. The orbit thus obtained is of a simple character but it possesses one important property; relative to the moving axes it is closed and the body following it will always return to the same point of it (relative to the moving axis) after the lapse of a definite interval. In other words, the relative motion is periodic.

Hill develops this solution literally and numerically for the case of our satellite with high accuracy. This accuracy is useful because the form of the orbit depends solely on the ratio of the mean rates of motion of the sun and moon round the earth, and these rates, determined from centuries of observation, are not affected by the various limitations imposed at the outset. The curve does not differ much from a circle to the eye but it includes the principal part of one of the chief differences of the motion from that in a circle with uniform velocity, namely, the inequality long known as the "variation"; hence the name since given to it, "the Variational Orbit." Hill,

however, saw that it was of more general interest than its particular application to our satellite. He proceeds to determine its form for other values
of the mean rates of motion of the two bodies. This gives a family of
periodic orbits whose form gradually varies as the ratio is changed; the
greater the ratio, the more the curve differs from a circle.

It is this idea of Hill's that has so profoundly changed the whole outlook
of celestial mechanics. Poincaré took it up as the basis of his celebrated
prize essay of 1887 on the problem of three bodies and afterwards expanded
his work into the three volumes; *Les méthodes nouvelles de la Mécanique
Céleste.* His treatment throughout is highly theoretical. He shows that
there must be many families of periodic orbits even for specialised problems
in the case of three bodies, certain general properties are found, and much
information concerning them which is fundamental for future investigation
is obtained.

It is doubtful if Darwin had paid any special attention to Hill's work
on the moon for at least ten years after its appearance. All this time he
was busy with the origin of the moon and with tidal work. Adams had
published a brief *résumé* of his own work on lines similar to those of Hill
immediately after the memoirs of the latter appeared, but nothing further
on the subject came from his pen. The medal of the Royal Astronomical
Society was awarded to Hill in 1888, and Dr Glaisher's address on his work
contains an illuminating analysis of the methods employed and the ideas
which are put forward. Probably both Darwin and Adams had a considerable share in making the recommendation. Darwin often spoke of his
difficulties in assimilating the work of others off his own beat and possibly
this address started him thinking about the subject, for it was at his recommendation in the summer of 1888 that the writer took up the study of Hill's
papers. "They seem to be very good," he said, "but scarcely anyone knows
much about them."

He lectured on Hill's work for the first time in the Michaelmas Term
of 1893, and writes of his difficulties in following parts of them, more
particularly that on the Moon's Perigee which contains the development of
the infinite determinant. He concludes, "I can't get on with my own work
until these lectures are over—but Hill's papers are splendid." One of his
pupils on this occasion was Dr P. H. Cowell, now Director of the Nautical
Almanac office. The first paper of the latter was a direct result of these
lectures and it was followed later by a valuable series of memoirs in which
the constants of the lunar orbit and the coefficients of many of the periodic
terms were obtained with great precision. Soon after these lectures Darwin
started his own investigations on the subject. But they took a different
line. The applications to the motion of the moon were provided for and
Poincaré had gone to the foundations. Darwin felt, however, that the work of

the latter was far too abstract to satisfy those who, like himself, frequently needed more concrete results, either for application or for their own mental satisfaction. In discussing periodic orbits he set himself the task of tracing numbers of them in order, as far as possible, to get a more exact knowledge of the various families which Poincaré's work had shown must exist. Some of Hill's original limitations are dropped. Instead of taking a sun of infinite mass and at an infinite distance, he took a mass ten times that of the planet and at a finite distance from that body. The orbit of each round the other is circular and of uniform motion, the third body being still of infinitesimal mass. Any periodic orbit which may exist is grist to his mill whether it circulate about one body or both or neither.

Darwin saw little hope of getting any extensive results by solutions of the differential equations in harmonic series. It was obvious that the slowness of convergence or the divergence would render the work far too doubtful. He adopted therefore the tedious process of mechanical quadratures, starting at an arbitrary position on the x-axis with an arbitrary speed in a direction parallel to the y-axis. Tracing the orbit step-by-step, he again reaches the x-axis. If the final velocity there is perpendicular to the axis, the orbit is periodic. If not, he starts again with a different speed and traces another orbit. The process is continued, each new attempt being judged by the results of the previous orbits, until one is obtained which is periodic. The amount of labour involved is very great since the actual discovery of a periodic orbit generally involved the tracing of from three to five or even more non-periodic paths. Concerning one of the orbits he traced for his last paper on the subject, he writes: "You may judge of the work when I tell you that I determined 75 positions and each averaged ¾ hr. (allowing for correction of small mistakes—which sometimes is tedious). You will see that it is far from periodic....I have now got six orbits of this kind." And all this to try and find only one periodic orbit belonging to a class of whose existence he was quite doubtful.

Darwin's previous work on figures of equilibrium of rotating fluids made the question of the stability of the motion in these orbits a prominent factor in his mind. He considered it an essential part in their classification. To determine this property it was necessary, after a periodic orbit had been obtained, to find the effect of a small variation of the conditions. For this purpose, Hill's second paper of 1877, on the Perigee of the Moon, is used. After finding the variation orbit in his first paper, Hill makes a start towards a complete solution of his limited differential equations by finding an orbit, not periodic and differing slightly from the periodic orbit already obtained. The new portion, the difference between the two, when expressed as a sum of harmonic terms, contains an angle whose uniform rate of change, c, depends only on the constants of the periodic orbit. The principal portion of Hill's paper is devoted to the determination of c with great precision. For this purpose, the infinite determinant is introduced and expanded into infinite series, the principal parts of which are expressed by a finite number of well known functions; the operations Hill devised to achieve this have always called forth a tribute to his skill. Darwin uses this constant c in a different way. If it is real, the orbit is stable, if imaginary, unstable. In the latter case, it may be a pure imaginary or a complex number; hence the necessity for the two kinds of unstability.

In order to use Hill's method, Darwin is obliged to analyse a certain function of the coordinates in the periodic orbit into a Fourier series, and to obtain the desired accuracy a large number of terms must be included. For the discovery of c from the infinite determinant, he adopts a mode of expansion of his own better suited to the purpose in hand. But in any case the calculation is laborious. In a later paper, he investigates the stability by a different method because Hill's method fails when the orbit has sharp flexures.

For the classification into families, Darwin follows the changes according to variations in the constant of relative energy, C. The differential equations referred to the moving axes admit a Jacobian integral, the constant of which is C. One property of this integral Hill had already developed, namely, that the curve obtained by making the kinetic energy zero is one which the body cannot cross. Darwin draws the curves for different values of C with care. He is able to show in several cases the origin of the families he has found and much use is made of Poincaré's proposition, that all such families originate in pairs, for following the changes. But even his material is sometimes insufficient, especially where two quite different pairs of families originate near the same point on the x-axis, and some later corrections of the classification partly by himself and partly by Mr S. S. Hough were necessary. In volume IV of his collected works these corrections are fully explained.

The long first memoir was published in 1896. Nothing further on the subject appeared from his hand until 1909 when a shorter paper containing a number of new orbits was printed in the Monthly Notices of the Royal Astronomical Society. Besides some additions and corrections to his older families he considers orbits of ejection and retrograde orbits. During the interval others had been at work on similar lines while Darwin with increasing duties thrust upon him only found occasional opportunities to keep his calculations going. A final paper which appears in the present volume was the outcome of a request by the writer that a trial should be made to find a member of a librating class of orbits for the mass ratio $1:10$ which had been shown to exist and had been traced for the mass ratio $1:1048$. The latter arose in an attempt to consider the orbits of the Trojan group of asteroids. He failed to find one but in the course of his work discovered another class of great interest, which shows the satellite ultimately falling into the planet. He concludes, "My attention was first drawn to periodic orbits by the desire to discover how a Laplacian ring could coalesce into a planet. With this object in view I tried to discover how a large planet could affect the mean motion of a small one moving in a circular orbit at the same mean distance. After various failures the investigation drifted towards the work of Hill and Poincaré, so that the original point of view was quite lost and it is not even mentioned in my paper on 'Periodic Orbits.' It is of interest, to me at least, to find that the original aspect of the problem has emerged again." It is of even greater interest to one of his pupils to find that after more than twenty years of work on different lines in celestial mechanics, Darwin's last paper should be on the same part of the subject to which both had been drawn from quite different points of view.

Thus Darwin's work on what appeared to be a problem in celestial mechanics of a somewhat unpractical nature sprang after all from and finally tended towards the question which had occupied his thoughts nearly all his life, the genesis and evolution of the solar system.

INAUGURAL LECTURE

(DELIVERED AT CAMBRIDGE, IN 1883, ON ELECTION TO THE
PLUMIAN PROFESSORSHIP)

I PROPOSE to take advantage of the circumstance that this is the first of the lectures which I am to give, to say a few words on the Mathematical School of this University, and especially of the position of a professor in regard to teaching at the present time.

There are here a number of branches of scientific study to which there are attached laboratories, directed by professors, or by men who occupy the position and do the duties of professors, but do not receive their pay from, nor full recognition by, the University. Of these branches of science I have comparatively little to say.

You are of course aware of the enormous impulse which has been given to experimental science in Cambridge during the last ten years. It would indeed have been strange if the presence of such men as now stand at the head of those departments had not created important Schools of Science. And yet when we consider the strange constitution of our University, it may be wondered that they have been able to accomplish this. I suspect that there may be a considerable number of men who go through their University course, whose acquaintance with the scientific activity of the place is limited by the knowledge that there is a large building erected for some obscure purpose in the neighbourhood of the Corn Exchange. Is it possible that any student of Berlin should be heard to exclaim, "Helmholtz, who is Helmholtz?" And yet some years ago I happened to mention the name of one of the greatest living mathematicians, a professor in this University, in the presence of a first class man and fellow of his College, and he made just such an exclamation.

This general state of apathy to the very existence of science here has now almost vanished, but I do not think I have exaggerated what it was some years ago. Is not there a feeling of admiration called for for those, who by their energy and ability have raised up all the activity which we now see?

For example, Foster arrived here, a stranger to the University, without University post or laboratory. I believe that during his first term Balfour and one other formed his whole class. And yet holding only that position of a College lecturer which he holds at this minute, he has come to make Cambridge the first Physiological School of Great Britain, and the range of buildings which the University has put at his disposal has already proved too small for his requirements*. His pupil Balfour had perhaps a less uphill game to play, for the germs of the School of Natural Science were already laid when he began his work as a teacher. But he did not merely aid in the further developments of what he found, for he struck out in a new line—that line of study which his own original work has gone, I believe, a very long way to transform and even create. He did not live to see the full development of the important school and laboratory which he had founded. But thanks to his impulse it is now flourishing, and will doubtless prosper under the able hands into which the direction has fallen. His name ought surely to live amongst us for what he did; for those who had the fortune to be his friends the remembrance of him cannot die, for what he was.

I should be going too far astray were I to continue to expatiate on the work of Rayleigh, Stuart, and the others who are carrying on the development of practical work in various branches within these buildings. It must suffice to say that each school has had its difficulties, and that those difficulties have been overcome by the zeal of those concerned in the management.

But now let us turn to the case of the scientific professors who have no laboratories to direct, and I speak now of the mathematical professors. In comparison with the prosperity of which I have been speaking, I think it is not too much to say that there is no vitality. I belong to this class of professors, and I am far from flattering myself that I can do much to impart life to the system. But if I shall not succeed I may perhaps be pardoned if I comfort myself by the reflection, that it may not be entirely my own fault.

The University has however just entered on a new phase; I have the honour to be the first professor elected under the new Statutes now in force. A new scheme for the examinations in Mathematics is in operation, and it may be that such an opportunity will now be afforded as has hitherto been wanting. We can but try to avail ourselves of the chance.

To what causes are we to assign the fact that our most eminent teachers of mathematics have hitherto been very frequently almost without classes? It surely cannot be that Stokes, Adams and Cayley have *nothing* to say worth hearing by students of mathematics. Granting the possibility

---

* Sir Michael Foster was elected the first Professor of Physiology a few weeks after the delivery of this lecture.

that a distinguished man may lack the power of exposition, yet it is inadmissible that they are *all* deficient in that respect. No, the cause is not far to seek, it lies in the Mathematical Tripos. How far it is desirable that the system should be so changed, that it will be advisable for students in their own interest to attend professorial lectures, I am not certain; but it can scarcely be doubted that if there were no Tripos, the attendance at such lectures would be larger.

In hearing the remarks which I am about to make on the Mathematical Tripos, you must bear in mind that I have hitherto taken no part in mathematical teaching of any kind, and therefore must necessarily be a bad judge of the possibilities of mathematical training, and of its effects on most minds. A year and a half ago I took part as Additional Examiner in the Mathematical Tripos, and I must confess that I was a good deal discouraged by what I saw. Now do not imagine that I flatter myself I was one jot better in all these respects than others, when I went through the mill. I too felt the pressure of time, and scribbled down all I could in my three hours, and doubtless presented to my examiners some very pretty muddles. I can only congratulate myself that the men I examined were not my competitors.

In order to determine whether anything can be done to improve this state of things, let us consider the merits and demerits of our Mathematical School. One of the most prominent evils is that our system of examination has a strong tendency to make men regard the subjects more as a series of isolated propositions than as a whole; and much attention has to be paid to a point, which is really important for the examination, viz. where to begin and where to leave off in answering a question. The *coup d'œil* of the whole subject is much impaired; but this is to some extent inherent in any system of examination. This result is, however, principally due to our custom of setting the examinees to reproduce certain portions of the books which they have studied; that is to say this evil arises from the "bookwork" questions. I have a strong feeling that such questions should be largely curtailed, and that the examinees should by preference be asked for transformations and modifications of the results obtained in the books. I suppose a certain amount of bookwork must be retained in order to permit patient workers, who are not favoured by any mathematical ability, to exhibit to the examiners that they have done their best. But for men with any mathematical power there can be no doubt that such questions as I suggest would give a far more searching test, and their knowledge of the subject would not have to be acquired in short patches.

I should myself like to see an examination in which the examinees were allowed to take in with them any books they required, so that they need not load their memories with formulae, which no original worker thinks of trying to remember. A first step in this direction has been taken by the introduction of logarithm tables into the Senate House; and I fancy that a terrible amount of incompetence was exhibited in the result. I may remark by the way that the art of computation is utterly untaught here, and that readiness with figures is very useful in ordinary life. I have done a good deal of such work myself, but I had to learn it by practice and from a few useful hints from others who had mastered it.

It is to be regretted that questions should be set in examinations which are in fact mere conjuring tricks with symbols, a kind of double acrostic; another objectionable class of question is the so-called physical question which has no relation to actual physics. This kind of question was parodied once by reference to "a very small elephant, whose weight may be neglected, etc." Examiners have often hard work to find good questions, and their difficulties are evidenced by such problems as I refer to. I think, however, that of late this kind of exercise is much less frequent than formerly.

I am afraid the impression is produced in the minds of many, that if a problem cannot be solved in a few hours, it cannot be solved at all. At any rate there seems to be no adequate realisation of the process by which most original work is done, when a man keeps a problem before him for weeks, months, years and gnaws away from time to time when any new light may strike him.

I think some of our text books are to blame in this; they impress the student in the same way that a high road must appear to a horse with blinkers. The road stretches before him all finished and macadamised, having existed for all he knows from all eternity, and he sees nothing of by-ways and foot-paths. Now it is the fact that scarcely any subject is so way worn that there are not numerous unexplored by-paths, which may lead across to undiscovered countries. I do not advocate that the student should be led along and made to examine all the cul-de-sacs and blind alleys, as he goes; he would never get on if he did so, but I do protest against that tone which I notice in many text books that mathematics is a spontaneously growing fruit of the tree of knowledge, and that all the fruits along *that* road have been gathered years ago. Rather let him see that the whole grand work is the result of the labours of an army of men, each exploring his little bit, and that there are acres of untouched ground, where he too may gather fruit: true, if he begins on original work, he may think that he has discovered something new and may very likely find that someone has been before him; but at least he *too* will have had the enormous pleasure of discovery.

There is another fault in the system of examinations, but I hardly know whether it can be appreciably improved. It is this:—the system gives very little training in the really important problem both of practical life and of mathematics, viz. the determination of the exact nature of the question which is to be attacked, the making up of your mind as to what you will do. Everyone who has done original work knows that at first the subject generally presents itself as a chaos of possible problems, and careful analysis is necessary before that chaos is disentangled. The process is exactly that of a barrister with his brief. A pile of papers is set before him, and from that pile he has to extract the precise question of law or fact on which the whole turns. When he has mastered the story and the precise point, he has generally done the more difficult part of his work. In most cases, it is exactly the same in mathematical work; and when the question has been pared down until its characteristics are those of a Tripos question, of however portentous a size, the battle is half won. It only remains to the investigator then to avail himself of all the "morbid aptitude for the manipulation of symbols" which he may happen to possess.

In examination, however, the whole of this preparatory part of the work is done by the examiner, and every examiner must call to mind the weary threshing of the air which he has gone through in trying "to get a question" out of a general idea. Now the limitation of time in an examination makes this evil to a large extent irremediable; but it seems to me that some good may be done by requesting men to write essays on particular topics, because in this case their minds are not guided by a pair of rails carefully prepared by an examiner.

In the report on the Tripos for 1882, I spoke of the slovenliness of style which characterised most of the answers. It appears to me that this is really much more than a mere question of untidiness and annoyance to examiners. The training here seems to be that form and style are matters of no moment, and answers are accordingly sent up in examination which are little more than rough notes of solutions. But I insist that a mathematical writer should attend to style as much as a literary man.

Some of our Cambridge writers on mathematics seem never to have recovered from the ill effects of their early training, even when they devote the rest of their life to original work. I wish some of you would look at the artistic mode of presentation practised by Gauss, and compare it with the standard of excellence which passes muster here. Such a comparison will not prove gratifying to our national pride.

Where there is slovenliness of style it is, I think, almost certain that there will be wanting that minute attention to form on which the successful, or at least easy, marshalling of a complex analytical development depends. The art of carrying out such work has to be learnt by trial and error by the men trained in our school, and yet the inculcation of a few maxims would generally be of great service to students, provided they are made to attend to them in their work. The following maxims contain the pith of the matter, although they might be amplified with advantage if I were to detain you over this point for some time.

1st. Choose the notation with great care, and where possible use a standard notation.

2nd. Break up the analysis into a series of subsections, each of which may be attended to in detail.

3rd. Never attempt too many transformations in one operation.

4th. Write neatly and not quickly, so that in passing from step to step there may be no mistakes of copying.

A man who undertakes any piece of work, and does not attend to some such rules as these, doubles his chances of mistake; even to short pieces of work such as examination questions the same applies, and I have little doubt that many a score of questions have been wrongly worked out from want of attention to these points.

It is true that great mathematicians have done their work in very various styles, but we may be sure that those who worked untidily gave themselves much unnecessary trouble. Within my own knowledge I may say that Thomson [Lord Kelvin] works in a copy-book, which is produced at Railway Stations and other conveniently quiet places for studious pursuits; Maxwell worked in part on the backs of envelopes and loose sheets of paper crumpled up in his pocket*; Adams' manuscript is as much a model of neatness in mathematical writing as Porson's of Greek writing. There is, of course, no infallibility in good writing, but believe me that untidiness surely has its reward in mistakes. I have spoken only on the evils of slovenliness in its bearing on the men as mathematicians—I cannot doubt that as a matter of general education it is deleterious.

I have dwelt long on the demerits of our scheme, because there is hope of amending some of them, but of the merits there is less to be said because they are already present. The great merit of our plan seems to me to be reaped only by the very ablest men in the year. It is that the student is enabled to get a wide view over a great extent of mathematical country, and if he has not assimilated all his knowledge thoroughly, yet he knows that it is so, and he has a fair introduction to many subjects. This advantage he would have lost had he become a pure specialist and original investigator very early in his career. But this advantage is all a matter of degree, and even the ablest man cannot cover an indefinitely long course

* I think that he must have been only saved from error by his almost miraculous physical insight, and by a knowledge of the time when work must be done neatly. But his *Electricity* was crowded with errata, which have now been weeded out one by one.

in his three years. Year by year new subjects were being added to the curriculum, and the limit seemed to have been exceeded; whilst the disastrous effects on the weaker brethren were becoming more prominent. I cannot but think that the new plan, by which a man shall be induced to become a partial specialist, gives us better prospects.

Another advantage we gain by our strict competition is that a man must be bright and quick; he must not sit mooning over his papers; he is quickly brought to the test,—either he can or he cannot do a definite problem in a finite time—if he cannot he is found out. Then if our scheme checks original investigation, it at least spares us a good many of those pests of science, the man who churns out page after page of x, y, z, and thinks he has done something in producing a mass of froth. That sort of man is quickly found out here, both for his own good and the good of the world at large. Lastly this place has the advantage of having been the training school of nearly all the English mathematicians of eminence, and of having always attracted—as it continues to attract—whatever of mathematical ability is to be found in the country. These are great merits, and in the endeavour to remove blemishes, we must see that we do not destroy them.

A discussion of the Mathematical Tripos naturally brings us face to face with a much abused word, namely "Cram."

The word connotes bad teaching, and accordingly teaching with reference to examinations has been supposed to be bad because it has been called cram. The whole system of private tuition commonly called coaching has been nick-named cram, and condemned accordingly. I can only say for myself that I went to a private tutor whose name is familiar to everyone in Cambridge, and found the most excellent and thorough teaching; far be it from me to pretend that I shall prove his equal as a teacher. Whatever fault is to be found, it is not with the teaching, but it lies in the system. It is obviously necessary that when a vast number of new subjects are to be mastered the most rigorous economy in the partition of the student's time must be practised, and he is on no account to be allowed to spend more than the requisite minimum on any one subject, even if it proves attractive to him. The private tutor must clearly, under the old regime, act as director of studies for his pupils strictly in accordance with examination requirements; for place in the Tripos meant pounds, shillings, and pence to the pupil. The system is now a good deal changed, and we may hope that it will be possible henceforth to keep the examination less incessantly before the student, who may thus become a student of a subject, instead of a student for a Tripos.

And now I think you must see the peculiar difficulties of a professor of mathematics; his vice has been that he tried to teach a subject *only*, and private tutors felt, and felt justly, that they could not, in justice to their pupils' prospects, conscientiously recommend the attendance at more than a very small number of professorial lectures. . But we are now at the beginning of a new regime and it may be that now the professors have their chance.   But I think it depends much more on the examiners than on the professors.   If examiners can and will conduct the examinations in such a manner that it shall "pay" better to master something thoroughly, than to have a smattering of much, we shall see a change in the manner of learning.   Otherwise there will not be much change.   I do not know how it will turn out, but I do know that it is the duty of professors to take such a chance if it exists.

My purpose is to try my best to lecture in such a way as will impart an interest to the subject itself and to help those who wish to learn, so that they may reap advantage in examinations—provided the examinations are conducted wisely.

INTRODUCTION TO DYNAMICAL ASTRONOMY

THE field of dynamical astronomy is a wide one and it is obvious that it will be impossible to consider even in the most elementary manner all branches of it; for it embraces all those effects in the heavens which may be attributed to the effects of gravitation. In the most extended sense of the term it may be held to include theories of gravitation itself. Whether or not gravitation is an ultimate fact beyond which we shall never penetrate is as yet unknown, but Newton, whose insight into physical causation was almost preternatural, regarded it as certain that some further explanation was ultimately attainable. At any rate from the time of Newton down to to-day men have always been striving towards such explanation—it must be admitted without much success. The earliest theory of the kind was that of Lesage, promulgated some 170 years ago. He conceived all space to be filled with what he called ultramundane corpuscles, moving with very great velocities in all directions. They were so minute and so sparsely distributed that their mutual collisions were of extreme rarity, whilst they bombarded the grosser molecules of ordinary matter. Each molecule formed a partial shield to its neighbours, and this shielding action was held to furnish an explanation of the mutual attraction according to the law of the inverse square of the distance, and the product of the areas of the sections of the two molecules. Unfortunately for this theory it is necessary to assume that there is a loss of energy at each collision, and accordingly there must be a perpetual creation of kinetic energy of the motion of the ultramundane corpuscles at infinity. The theory is further complicated by the fact that the energy lost by the corpuscle at each collision must have been communicated to the molecule of matter, and this must occur at such a rate as to vaporize all matter in a small fraction of a second. Lord Kelvin has, however, pointed out that there is a way out of this fundamental difficulty, for if at each collision the ultramundane corpuscle should suffer no loss of total kinetic energy but only a transformation of energy of translation into energy of internal vibration, the system becomes conservative of energy and the eternal creation of energy becomes unnecessary. On the other hand, gravitation will not be transmitted to infinity, but only to a limited distance.

I will not refer further to this conception save to say that I believe that no man of science is disposed to accept it as affording the true road.

It may be proved that if space were an absolute plenum of incompressible fluid, and that if in that fluid there were points towards which the fluid streams from all sides and disappears, those points would be urged towards one another with a force varying inversely as the square of the distance and directly as the product of the intensities of the two inward streams. Such points are called sinks and the converse, namely points from whence the fluid streams, are called sources. Now two sources also attract one another according to the same law; on the other hand a source and a sink repel one another. If we could conceive matter to be all sources or all sinks we should have a mechanical theory of gravitation, but no one has as yet suggested any means by which this can be realised. Bjerknes of Christiania has, however, suggested a mechanical means whereby something of the kind may be realised. Imagine an elastic ball immersed in water to swell and contract rhythmically, then whilst it is contracting the motion of the surrounding water is the same as that due to a sink at its centre, and whilst it is expanding the motion is that due to a source. Hence two balls which expand and contract in exactly the same phase will attract according to the law of gravitation on taking the average over a period of oscillation. If, however, the pulsations are in opposite phases the resulting force is one of repulsion. If then all matter should resemble in some way the pulsating balls we should have an explanation, but the absolute synchronism of the pulsations throughout all space imports a condition which does not commend itself to physicists. I may mention that Bjerknes has actually realised these conclusions by experiment. Although it is somewhat outside our subject I may say that if a ball of invariable volume should execute a small rectilinear oscillation, its advancing half gives rise to a source and the receding half to a sink, so that the result is what is called a doublet. Two oscillating balls will then exercise on one another forces analogous to that of magnetic particles, but the forces of magnetism are curiously inverted. This quasi-magnetism of oscillating balls has also been treated experimentally by Bjerknes. However curious and interesting these speculations and experiments may be, I do not think they can afford a working hypothesis of gravitation.

A new theory of gravitation which appears to be one of extraordinary ingenuity has lately been suggested by a man of great power, viz. Osborne Reynolds, but I do not understand it sufficiently to do more than point out the direction towards which he tends. He postulates a molecular ether. I conceive that the molecules of ether are all in oscillation describing orbits in the neighbourhood of a given place. If the region of each molecule be replaced by a sphere those spheres may be packed in a hexagonal arrangement completely filling all space. We may, however, come to places where the symmetrical piling is interrupted, and Reynolds calls this a region of misfit.

Then, according to this theory, matter consists of misfit, so that matter is the deficiency of molecules of ether. Reynolds claims to show that whilst the particular molecules which don't fit are continually changing the amount of misfit is indestructible, and that two misfits attract one another. The theory is also said to explain electricity. Notwithstanding that Reynolds is not a good exponent of his own views, his great achievements in science are such that the theory must demand the closest scrutiny.

The newer theories of electricity with which the name of Prof. J. J. Thomson is associated indicate the possibility that mass is merely an electro-dynamic phenomenon. This view will perhaps necessitate a revision of all our accepted laws of dynamics. At any rate it will be singular if we shall have to regard electrodynamics as the fundamental science, and subsequently descend from it to the ordinary laws of motion. How much these notions are in the air is shown by the fact that at a congress of astronomers, held in 1902 at Göttingen, the greater part of one day's discussion was devoted to the astronomical results which would follow from the new theory of electrons.

I have perhaps said too much about the theories of gravitation, but it should be of interest to you to learn how it teems with possibilities and how great is the present obscurity.

Another important subject which has an intimate relationship with Dynamical Astronomy is that of abstract dynamics. This includes the general principles involved in systems in motion under the action of con-servative forces and the laws which govern the stability of systems. Perhaps the most important investigators in this field are Lagrange and Hamilton, and in more recent times Lord Kelvin and Poincaré.

Two leading divisions of dynamical astronomy are the planetary theory and the theory of the motion of the moon and of other satellites. A first approximation in all these cases is afforded by the case of simple elliptic motion, and if we are to consider the case of comets we must include parabolic and hyperbolic motion round a centre. Such a first approximation is, however, insufficient for the prediction of the positions of any of the bodies in our solar system for any great length of time, and it becomes necessary to include the effects of the disturbing action of one or more other bodies. The problem of disturbed revolution may be regarded as a single problem in all its cases, but the defects of our analysis are such that in effect its several branches become very distinct from one another. It is usual to speak of the problem of disturbed revolution as the problem of three bodies, for if it were possible to solve the case where there are three bodies we should already have gone a long way towards the solution of that more complex case where there are any number of bodies.

Owing to the defects of our analysis it is at present only possible to obtain accurate results of a general character by means of tedious expansions. All the planets and all the satellites have their motions represented with more or less accuracy by ellipses, but this first approximation ceases to be satisfactory for satellites much more rapidly than is the case for planets. The eccentricities of the ellipses and the inclinations of the orbits are in most cases inconsiderable. It is assumed then that it is possible to effect the requisite expansions in powers of the eccentricities and of suitable functions of the inclinations. Further than this it is found necessary to expand in powers of the ratios of the mean distances of the disturbed and disturbing bodies from the centre. It is at this point that the first marked separation of the lunar and planetary theories takes place. In the lunar theory the distance of the sun (disturber) from the earth is very great compared with that of the moon, and we naturally expand in this ratio in order to start with as few terms as possible. In the planetary theory the ratio of the distances of the disturbed and disturbing bodies—two planets—from the sun may be a large fraction. For example, the mean distances of Venus and the earth are approximately in the ratio 7 : 10, and in order to secure sufficient accuracy a large number of terms is needed. In the case of the planetary theory the expansion is delayed as long as possible.

Again, in the lunar theory the mass of the disturbing body is very great compared with that of the primary, a ratio on which it is evident that the amount of perturbation greatly depends. On the other hand, in the planetary theory the disturbing body has a very small mass compared with that of the primary, the sun. From these facts we are led to expect that large terms will be present in the expressions for the motion of the moon due to the action of the sun, and that the later terms in the expansion will rapidly decrease; and in the planetary theory we expect large numbers of terms all of about equal magnitude and none of them very great. This expectation is, however, largely modified by some further remarks to be made.

You know that a dynamical system may have various modes of free oscillation of various periods. If then a disturbing force with a period differing but little from that of one of the modes of free oscillation acts on the system for a long time it will generate an oscillation of large amplitude.

A familiar instance of this is in the roll of a ship at sea. If the incidence of the waves on the ship is such that the succession of impulses is very nearly identical in period with the natural period of the ship, the roll becomes large. In analysis this physical fact is associated with a division by a small divisor on integration.

As an illustration of the simplest kind suppose that the equation of motion of a system under no forces were

$$\frac{d^2x}{dt^2} + n^2x = 0.$$

Then we know that the solution is

$$x = A \cos nt + B \sin nt,$$

that is to say the free period is $\frac{2\pi}{n}$. Suppose then such a system be acted on by a perturbing force $F \cos (n - \epsilon) t$, where ϵ is small; the equation of motion is

$$\frac{d^2x}{dt^2} + n^2x = F \cos (n - \epsilon) t,$$

and the solution corresponding to such a disturbing force is

$$x = \frac{F}{-(n - \epsilon)^2 + n^2} \cos (n - \epsilon) t = \frac{F}{2n\epsilon - \epsilon^2} \cos (n - \epsilon) t.$$

If ϵ is small the amplitude becomes great, and this arises, as has been said, by a division by a small divisor.

Now in both lunar and planetary theories the coefficients of the periodic terms become frequently much greater than might have been expected *à priori*. In the lunar theory before this can happen in such a way as to cause much trouble the coefficients have previously become so small that it is not necessary to consider them. But suppose in the planetary theory n, n' are the mean motions of two planets round the primary. Then coefficients will continually be having multipliers of the forms

$$\frac{n'}{in \pm i'n'} \text{ and } \left(\frac{n'}{in \pm i'n'}\right)^2,$$

where i, i' are small positive integers. In general the larger i, i' the smaller is the coefficient to begin with, but owing to the fact that the ratio $n : n'$ may very nearly approach that of two small integers a coefficient may become very great; e.g. 5 Jovian years nearly equal 2 of Saturn, while the ratio of the mean distances is 6 : 11. The result is a large long inequality with a period of 913 years in the motions of those two planets. The periods of the principal terms in the moon's motion are generally short, but some have large coefficients, so that the deviation from elliptic motion is well marked.

The general problem of three bodies is in its infancy, and as yet but little is known as to the possibilities in the way of orbits and as to their stabilities.

Another branch of our subject is afforded by the precession and nutation of the earth, or any other planet, under the influence of the attractions of disturbing bodies. This is the problem of disturbed rotation and it presents a strong analogy with the problem of disturbed elliptic motion. When a top spins with absolute steadiness we say that it is asleep. Now the earth in its rotation may be asleep or it may not be so—there is nothing but observation which is capable of deciding whether it is so or not. This is equally true whether the rotation takes place under external perturbation or not. If the earth is asleep its motion presents a perfect analogy with circular orbital motion; if it wobbles the analogy is with elliptic motion. The analogy is such that the magnitude of the wobble corresponds with the eccentricity of orbit and the position of greatest departure with the longitude of pericentre. Until the last 20 years it has always been supposed that the earth is asleep in its rotation, but the extreme accuracy of modern observation, when subjected to the most searching analysis by Chandler and others, has shewn that there is actually a small wobble. This is such that the earth's axis of rotation describes a small circle about the pole of figure. The theory of precession indicated that this circle should be described in a period of 305 days, and all the earlier astronomers scrutinised the observations with the view of detecting such an inequality. It was this preconception, apparently well founded, which prevented the detection of the small inequality in question. It was Chandler who first searched for an inequality of unknown period and found a clearly marked one with a period of 428 days. He found also other smaller inequalities with a period of a year. This wandering of the pole betrays itself most easily to the observer by changes in the latitude of the place of observation.

The leading period in the inequality of latitude is then one of 428 days. The theoretical period of 305 days was, as I have said, apparently well established, but after the actual period was found to be 428 days Newcomb pointed out that if the earth is not absolutely rigid, but slightly changes its shape as the axis of rotation wanders, such a prolongation of period would result. Thus these purely astronomical observations end by affording a measure of the effective rigidity of the earth's mass.

The theory of the earth's figure and the variation of gravity as we vary our position on the surface or the law of variation of gravity as we descend into mines are to be classified as branches of dynamical astronomy, although in these cases the velocities happen to be zero. This theory is intimately connected with that of precession, for it is from this that we conclude that the free wobble of the perfectly rigid earth should have a period of 305 days. The ellipticity of the earth's figure also has an important influence on the motion of the moon, and the determination of a certain inequality in the moon's motion affords the means of finding the amount of ellipticity of the earth's figure with perhaps as great an accuracy as by any other means. Indeed in the case of Jupiter, Saturn, Mars, Uranus and Neptune the ellipticity is most accurately determined in this way. The masses also of the planets may be best determined by the periods of their satellites.

The theory of Saturn's rings is another branch. The older and now obsolete views that the rings are solid or liquid gave the subject various curious and difficult mathematical investigations. The modern view—now well established—that they consist of an indefinite number of meteorites which collide together from time to time presents a number of problems of great difficulty. These were ably treated by Maxwell, and there does not seem any immediate prospect of further extension in this direction.

Then the theory of the tides is linked to astronomy through the fact that it is the moon and sun which cause the tides, so that any inequality in their motions is reflected in the ocean.

On the fringe of our subject lies the whole theory of figures of equilibrium of rotating liquids with the discussion of the stability of the various possible forms and the theory of the equilibrium of gaseous planets. In this field there is yet much to discover.

This subject leads on immediately to theories of the origin of planetary systems and to cosmogony. Tidal theory, on the hypothesis that the tides are resisted by friction, leads to a whole series of investigations in speculative astronomy whose applications to cosmogony are of great interest.

Up to a recent date there was little evidence that gravitation held good outside the solar system, but recent investigations, carried out largely by means of the spectroscopic determinations of velocities of stars in the line of sight, have shewn that there are many other systems, differing very widely from our own, where the motions seem to be susceptible of perfect explanation by the theory of gravitation. These new extensions of gravitation outside our system are leading to many new problems of great difficulty and we may hope in time to acquire wider views as to the possibilities of motion in the heavens.

This hurried sketch of our subject will shew how vast it is, and I cannot hope in these lectures to do more than touch on some of the leading topics.

HILL'S LUNAR THEORY

§ 1. *Introduction**.

AN account of Hill's *Lunar Theory* can best be prefaced by a few quotations from Hill's original papers. These will indicate the peculiarities which mark off his treatment from that of earlier writers and also, to some extent, the reasons for the changes he introduced. Referring to the well-known expressions which give, for undisturbed elliptic motion, the rectangular coordinates as explicit functions of the time—expressions involving nothing more complicated than Bessel's functions of integral order—Hill writes:

"Here the law of series is manifest, and the approximation can easily be carried as far as we wish. But the longitude and latitude, variables employed by nearly all lunar theorists, are far from having such simple expressions; in fact their coefficients cannot be finitely expressed in terms of Besselian functions. And if this is true in the elliptic theory how much more likely is a similar thing to be true when the complexity of the problem is increased by the consideration of disturbing forces?...There is also another advantage in employing coordinates of the former kind (rectangular): the differential equations are expressed in purely algebraic functions, while with the latter (polar) circular functions immediately present themselves."

In connection with the parameters to be used in the expansions Hill argues thus:

"Again as to parameters all those who have given literal developments, Laplace setting the example, have used the parameter *m*, the ratio of the sidereal month to the sidereal year. But a slight examination, even of the results obtained, ought to convince anyone that this is a most unfortunate selection in regard to convergence. Yet nothing seems to render the parameter desirable, indeed the ratio of the synodic month to the sidereal year would appear to be more naturally suggested as a parameter."

---

* The references in this section are to Hill's "Researches in the Lunar Theory" first published (1878) in the *American Journal of Mathematics*, vol. I. pp. 5—26, 129—147 and reprinted in *Collected Mathematical Works*, vol. I. pp. 284—335. Hill's other paper connected with these lectures is entitled "On the Part of the Motion of the Lunar Perigee which is a function of the Mean Motions of the Sun and Moon," published separately in 1877 by John Wilson and Son, Cambridge, Mass., and reprinted in *Acta Mathematica*, vol. VIII. pp. 1—36, 1886 and in *Collected Mathematical Works*, vol. I. pp. 243—270.

When considering the order of the differential equations and the method of integration, Hill wrote:

" Again the method of integration by undetermined coefficients is most likely to give us the nearest approach to the law of series; and in this method it is as easy to integrate a differential equation of the second order as one of the first, while the labour is increased by augmenting the number of variables and equations. But Delaunay's method doubles the number of variables in order that the differential equations may be all of the first order. Hence in this disquisition I have preferred to use the equations expressed in terms of the coordinates rather than those in terms of the elements; and, in general, always to diminish the number of unknown quantities and equations by augmenting the order of the latter. In this way the labour of making a preliminary development of R in terms of the elliptic elements is avoided."

We may therefore note the characteristics of Hill's method as follows:

(1) Use of rectangular coordinates.

(2) Expansion of series in powers of the ratio of the synodic month to the sidereal year.

(3) Use of differential equations of the second order which are solved by assuming series of a definite type and equating coefficients.

In these lectures we shall obtain only the first approximation to the solution of Hill's differential equations. The method here followed is not that given by Hill, although it is based on the same principles as his method. Our work only involves simple algebra, and probably will be more easily understood than Hill's. If followed in detail to further approximations, it would prove rather tedious, but it leads to the results we require without too much labour. If it is desired to follow out the method further, reference should be made to Hill's own writings.

§ 2. Differential Equations of Motion and Jacobi's Integral.

Let E, M, m' denote the masses or positions of the earth, moon, and sun, and let G be the centre of inertia of E and M. Let x, y, z be the rectangular coordinates of M with E as origin, and let x', y', z' be the coordinates of m' referred to parallel axes through G. The coordinates of M relative to the axes through G are clearly $\dfrac{E}{E+M}x$, $\dfrac{E}{E+M}y$, $\dfrac{E}{E+M}z$; those of E are $-\dfrac{M}{E+M}x$, $-\dfrac{M}{E+M}y$, $-\dfrac{M}{E+M}z$. The distances $\overline{EM}$, $\overline{Em'}$, $\overline{Mm'}$ are denoted by r, r_1, Δ respectively. It is assumed that G describes a Keplerian ellipse round m' so that x', y', z' are known functions of the time. The accelerations of M relative to E are shewn in the diagram.

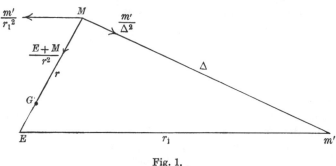

Fig. 1.

We have
$$r^2 = x^2 + y^2 + z^2,$$

$$r_1{}^2 = \left(x' + \frac{Mx}{E+M}\right)^2 + \left(y' + \frac{My}{E+M}\right)^2 + \left(z' + \frac{Mz}{E+M}\right)^2,$$

$$\Delta^2 = \left(x' - \frac{Ex}{E+M}\right)^2 + \left(y' - \frac{Ey}{E+M}\right)^2 + \left(z' - \frac{Ez}{E+M}\right)^2.$$

Hence
$$\frac{\partial r}{\partial x} = \frac{x}{r},$$

$$\frac{E+M}{M}\frac{\partial r_1}{\partial x} = \frac{x' + \dfrac{Mx}{E+M}}{r_1},$$

$$-\frac{E+M}{E}\frac{\partial \Delta}{\partial x} = \frac{x' - \dfrac{Ex}{E+M}}{\Delta};$$

$\therefore$ the direction cosines of EM are $\dfrac{\partial r}{\partial x}$, $\dfrac{\partial r}{\partial y}$, $\dfrac{\partial r}{\partial z}$,

    ,,   ,,   ,, Em' are $\dfrac{E+M}{M}\left(\dfrac{\partial r_1}{\partial x},\ \dfrac{\partial r_1}{\partial y},\ \dfrac{\partial r_1}{\partial z}\right)$,

    ,,   ,,   ,, Mm' are $-\dfrac{E+M}{E}\left(\dfrac{\partial \Delta}{\partial x},\ \dfrac{\partial \Delta}{\partial y},\ \dfrac{\partial \Delta}{\partial z}\right).$

If X, Y, Z denote the components of acceleration of M relative to axes through E,

$$X = -\frac{E+M}{r^2}\frac{\partial r}{\partial x} - \frac{m'}{\Delta^2}\frac{E+M}{E}\frac{\partial \Delta}{\partial x} - \frac{m'}{r_1^2}\frac{E+M}{M}\frac{\partial r_1}{\partial x}$$

$$= \frac{\partial F}{\partial x},$$

where $\qquad F = \dfrac{E+M}{r} + \dfrac{m'}{\Delta}\dfrac{E+M}{E} + \dfrac{m'}{r_1}\dfrac{E+M}{M}.$

$\qquad\qquad\qquad\qquad\qquad\qquad\qquad\qquad\qquad\qquad$ $\Big\}$ ......(1).

Similarly, $\quad Y = \dfrac{\partial F}{\partial y}, \quad Z = \dfrac{\partial F}{\partial z}.$

Let r' be the distance between G and m', and let θ be the angle $m'GM$; then

$$r'^2 = x'^2 + y'^2 + z'^2 \quad \text{and} \quad \cos\theta = \frac{xx' + yy' + zz'}{rr'},$$

$$r_1^2 = r'^2 + \frac{2M}{E+M}rr'\cos\theta + \left(\frac{Mr}{E+M}\right)^2,$$

$$\Delta^2 = r'^2 - \frac{2E}{E+M}rr'\cos\theta + \left(\frac{Er}{E+M}\right)^2.$$

Since r is very small compared with r',

$$\frac{1}{r_1} = \frac{1}{r'}\left\{1 - \frac{M}{E+M}\frac{r}{r'}\cos\theta + \left(\frac{M}{E+M}\cdot\frac{r}{r'}\right)^2 (\tfrac{3}{2}\cos^2\theta - \tfrac{1}{2})\dots\right\},$$

$$\frac{1}{\Delta} = \frac{1}{r'}\left\{1 + \frac{E}{E+M}\frac{r}{r'}\cos\theta + \left(\frac{E}{E+M}\cdot\frac{r}{r'}\right)^2 (\tfrac{3}{2}\cos^2\theta - \tfrac{1}{2})\dots\right\}.$$

$$\therefore \quad \frac{1}{E\Delta} + \frac{1}{Mr_1} = \frac{E+M}{EM}\cdot\frac{1}{r'} + \frac{1}{E+M}\frac{r^2}{r'^3}(\tfrac{3}{2}\cos^2\theta - \tfrac{1}{2})\dots.$$

Hence $\qquad F = \dfrac{E+M}{r} + \dfrac{m'(E+M)^2}{EMr'} + \dfrac{m'r^2}{r'^3}(\tfrac{3}{2}\cos^2\theta - \tfrac{1}{2})\dots.$

But the second term does not involve x, y, z, and may be dropped.

$$\therefore \quad F = \frac{E+M}{r} + \frac{m'r^2}{r'^3}(\tfrac{3}{2}\cos^2\theta - \tfrac{1}{2})\dots\dots\dots\dots\dots(2),$$

neglecting terms in $\dfrac{r^3}{r'^4}$.

We will now find an approximate expression for F', paying attention to the magnitude of the various terms in the actual earth-moon-sun system. As a first rough approximation, r' is a constant a', and Gm' rotates with uniform angular velocity n'. This neglects the effect on the sun of the earth and moon not being collected at G (this effect is very small), and it neglects the eccentricity of the solar orbit. In order that the coordinates of the sun relative to the earth might be nearly constant, we introduce axes x, y

rotating with angular velocity n' in the plane of the sun's orbit round the earth; the x-axis being so chosen that it passes through the sun. When required, a z-axis is taken perpendicular to the plane of x, y. As before, let x, y, z be the coordinates of the moon; the sun's coordinates will be approximately a', 0, 0. In this approximation $r \cos \theta = x$ and

$$F = \frac{E + M}{r} + \tfrac{3}{2} \frac{m'}{a'^3} x^2 - \tfrac{1}{2} m' \frac{r^2}{a'^3}.$$

This suggests the following general form for F, instead of that given in equation (2):

$$F = \frac{E + M}{r} + \tfrac{3}{2} \frac{m'}{a'^3} x^2 + \tfrac{3}{2} m' \left(\frac{r^2 \cos^2 \theta}{r'^3} - \frac{x^2}{a'^3} \right)$$
$$- \tfrac{1}{2} \frac{m'}{a'^3} (x^2 + y^2) - \tfrac{1}{2} \frac{m'}{a'^3} z^2 + \tfrac{1}{2} m' r^2 \left(\frac{1}{a'^3} - \frac{1}{r'^3} \right).$$

For the sake of future developments, we now introduce a new notation. Let ν be the moon's synodic mean motion and put $\mathrm{m} = \dfrac{n'}{\nu} = \dfrac{n'}{n - n'}$.* In the case of our moon, m is approximately $\frac{1}{12}$: this is a small quantity in powers of which our expressions will be obtained. If we neglect E and M compared with m', we have $m' = n'^2 a'^3$, whence $\dfrac{m'}{a'^3} = n'^2 = \nu^2 \mathrm{m}^2$. Let us also write $E + M = \kappa \nu^2$, and then we get

$$F + \tfrac{1}{2} n'^2 (x^2 + y^2) = \nu^2 \left[\frac{\kappa}{r} + \tfrac{1}{2} \mathrm{m}^2 (3x^2 - z^2) + \tfrac{3}{2} \mathrm{m}^2 \left(\frac{a'^3}{r'^3} r^2 \cos^2 \theta - x^2 \right) \right.$$
$$\left. + \tfrac{1}{2} \mathrm{m}^2 r^2 \left(1 - \frac{a'^3}{r'^3} \right) \right].$$

For convenience we write

$$\Omega = \tfrac{3}{2} \mathrm{m}^2 \left(\frac{a'^3}{r'^3} r^2 \cos^2 \theta - x^2 \right) + \tfrac{1}{2} \mathrm{m}^2 r^2 \left(1 - \frac{a'^3}{r'^3} \right),$$

and then $$F + \tfrac{1}{2} n'^2 (x^2 + y^2) = \nu^2 \left[\frac{\kappa}{r} + \tfrac{1}{2} \mathrm{m}^2 (3x^2 - z^2) + \Omega \right].$$

The equations of motion for uniformly rotating axes† are

$$\frac{d^2 x}{dt^2} - 2n' \frac{dy}{dt} - n'^2 x = \frac{\partial F}{\partial x}$$
$$\frac{d^2 y}{dt^2} + 2n' \frac{dx}{dt} - n'^2 y = \frac{\partial F}{\partial y}$$
$$\frac{d^2 z}{dt^2} \qquad\qquad = \frac{\partial F}{\partial z}$$

---

* In the lunar theory n' is supposed to be a known constant, while n (or m) is one of the constants of integration the value of which is not yet determined and can only be determined from the observations. So far n (or m) is quite arbitrary.

† See any standard treatise on Dynamics.

which give

$$\frac{d^2x}{dt^2} - 2n'\frac{dy}{dt} = \frac{\partial}{\partial x}\left[F + \tfrac{1}{2}n'^2(x^2 + y^2)\right] = \nu^2\left[-\frac{\kappa x}{r^3} + 3\mathrm{m}^2 x + \frac{\partial\Omega}{\partial x}\right],$$

$$\frac{d^2y}{dt^2} + 2n'\frac{dx}{dt} = \frac{\partial}{\partial y}\left[F + \tfrac{1}{2}n'^2(x^2 + y^2)\right] = \nu^2\left[-\frac{\kappa y}{r^3} \quad\quad + \frac{\partial\Omega}{\partial y}\right],$$

$$\frac{d^2z}{dt^2} \quad\quad = \frac{\partial}{\partial z}\left[F + \tfrac{1}{2}n'^2(x^2 + y^2)\right] = \nu^2\left[-\frac{\kappa z}{r^3} - \mathrm{m}^2 z + \frac{\partial\Omega}{\partial z}\right].$$

We might write $\tau = \nu t$, and on dividing the equations by ν^2 use τ henceforth as equivalent to time; or we might choose a special unit of time such that ν is unity. In either case our equations become

$$\left.\begin{aligned}
\frac{d^2x}{d\tau^2} - 2\mathrm{m}\frac{dy}{d\tau} + \frac{\kappa x}{r^3} - 3\mathrm{m}^2 x &= \frac{\partial\Omega}{\partial x} \\[1mm]
\frac{d^2y}{d\tau^2} + 2\mathrm{m}\frac{dx}{d\tau} + \frac{\kappa y}{r^3} \quad\quad &= \frac{\partial\Omega}{\partial y} \\[1mm]
\frac{d^2z}{d\tau^2} \quad\quad + \frac{\kappa z}{r^3} + \mathrm{m}^2 z &= \frac{\partial\Omega}{\partial z}
\end{aligned}\right\} \quad\dots\dots\dots\dots\dots(3).$$

If we multiply these equations respectively by $2\dfrac{dx}{d\tau}$, $2\dfrac{dy}{d\tau}$, $2\dfrac{dz}{d\tau}$ and add them, we have

$$\frac{d}{d\tau}\left\{\left(\frac{dx}{d\tau}\right)^2 + \left(\frac{dy}{d\tau}\right)^2 + \left(\frac{dz}{d\tau}\right)^2\right\} - 2\kappa\frac{d}{d\tau}\left(\frac{1}{r}\right) - 3\mathrm{m}^2\frac{d}{d\tau}(x^2) + \mathrm{m}^2\frac{d}{d\tau}(z^2)$$

$$= 2\left(\frac{\partial\Omega}{\partial x}\frac{dx}{d\tau} + \frac{\partial\Omega}{\partial y}\frac{dy}{d\tau} + \frac{\partial\Omega}{\partial z}\frac{dz}{d\tau}\right).$$

The whole of the left-hand side is a complete differential; the right-hand side needs the addition of the term $2\dfrac{\partial\Omega}{\partial\tau}$.

Let us put for brevity

$$V^2 = \left(\frac{dx}{d\tau}\right)^2 + \left(\frac{dy}{d\tau}\right)^2 + \left(\frac{dz}{d\tau}\right)^2.$$

Then

$$V^2 = \frac{2\kappa}{r} + 3\mathrm{m}^2 x^2 - \mathrm{m}^2 z^2 + 2\int_0^\tau \left[\frac{\partial\Omega}{\partial x}\frac{dx}{d\tau} + \frac{\partial\Omega}{\partial y}\frac{dy}{d\tau} + \frac{\partial\Omega}{\partial z}\frac{dz}{d\tau}\right]d\tau + C \dots(4).$$

If the earth moved round the sun with uniform angular velocity n', the axis of x would always pass through the sun, and therefore we should have

$$x' = r' = a', \quad y' = z' = 0$$

and

$$r\cos\theta = \frac{xx' + yy' + zz'}{r'} = x,$$

giving $$\frac{a'^3}{r'^3}\,r^2\cos^2\theta - x^2 = 0.$$

In this case Ω would vanish. It follows that Ω must involve as a factor the eccentricity of the solar orbit.

It is proposed as a first approximation to neglect that eccentricity, and this being the case, our equations become

$$\left.\begin{aligned}
&\frac{d^2x}{d\tau^2} - 2\mathrm{m}\frac{dy}{d\tau} + \frac{\kappa x}{r^3} - 3\mathrm{m}^2x = 0 \\[2mm]
&\frac{d^2y}{d\tau^2} + 2\mathrm{m}\frac{dx}{d\tau} + \frac{\kappa y}{r^3} \qquad\;\; = 0 \\[2mm]
&\frac{d^2z}{d\tau^2} \qquad\quad + \frac{\kappa z}{r^3} + \;\; \mathrm{m}^2z = 0
\end{aligned}\right\}\dots\dots\dots\dots(5).$$

Of these equations one integral is known, viz. Jacobi's integral,

$$V^2 = 2\,\frac{\kappa}{r} + 3\mathrm{m}^2x^2 - \mathrm{m}^2z^2 + C.$$

§ 3. *The Variational Curve.*

In ordinary theories the position of a satellite is determined by the departure from a simple ellipse—fixed or moving. The moving ellipse is preferred to the fixed one, because it is found that the departures of the actual body from the moving ellipse are almost of a periodic nature. But the moving ellipse is not the solution of any of the equations of motion occurring in the theory. Instead of referring the true orbit to an ellipse, Hill introduced as the orbit of reference, or intermediate orbit, a curve suggested by his differential equations, called the "variational curve."

We have already neglected the eccentricity of the solar orbit, and will now go one step further and neglect the inclination of the lunar orbit to the ecliptic, so that z disappears. If the path of a body whose motion satisfies

$$\left.\begin{aligned}
&\frac{d^2x}{d\tau^2} - 2\mathrm{m}\frac{dy}{d\tau} + \left(\frac{\kappa}{r^3} - 3\mathrm{m}^2\right)x = 0 \\[2mm]
&\frac{d^2y}{d\tau^2} + 2\mathrm{m}\frac{dx}{d\tau} + \frac{\kappa y}{r^3} \qquad = 0
\end{aligned}\right\}\dots\dots\dots\dots(6)$$

intersects the x-axis at right angles, the circumstances of the motion before and after intersection are identical, but in reverse order. Thus, if time be counted from the intersection, $x = f(\tau^2)$, $y = \tau f(\tau^2)$; for if in the differential equations the signs of y and τ are reversed, but x left unchanged, the equations are unchanged.

A similar result holds if the path intersects y at right angles, for if x and τ have signs changed, but y is unaltered, the equations are unaltered.

Now it is evident that the body may start from a given point on the x-axis, and at right angles to it, with different velocities, and that within certain limits it may reach the axis of y and cross it at correspondingly different angles. If the right angle lie between some of these, we judge from the principle of continuity that there is some intermediate velocity with which the body would arrive at and cross the y-axis at right angles.

If the body move from one axis to the other, crossing both at right angles, it is plain that the orbit is a closed curve symmetrical to both axes. Thus is obtained a particular solution of the differential equations. This solution is the "variational curve." While the general integrals involve four arbitrary constants, the variational curve has but two, which may be taken to be the distance from the origin at the x crossing and the time of crossing.

For the sake of brevity, we may measure time from the instant of crossing x.

Then since x is an even function of τ and y an odd one, both of period 2π, it must be possible to expand x and y by Fourier Series—thus

$$x = A_0 \cos \tau + A_1 \cos 3\tau + A_2 \cos 5\tau + \ldots\ldots,$$
$$y = B_0 \sin \tau + B_1 \sin 3\tau + B_2 \sin 5\tau + \ldots\ldots.$$

When τ is a multiple of π, $y = 0$; and when it is an odd multiple of $\dfrac{\pi}{2}$, $x = 0$: also in the first case $\dfrac{dx}{d\tau} = 0$ and in the second $\dfrac{dy}{d\tau} = 0$. Thus these conditions give us the kind of curve we want. It will be noted that there are no terms with even multiples of τ; such terms have to be omitted if $x, \dfrac{dy}{d\tau}$ are to vanish at $\tau = \pi/2$, etc.

We do not propose to follow Hill throughout the arduous analysis by which he determines the nature of this curve with the highest degree of accuracy, but will obtain only the first rough approximation to its form— thereby merely illustrating the principles involved.

Accordingly we shall neglect all terms higher than those in 3τ. It is also convenient to change the constants into another form. Thus we write

$$A_0 = a_0 + a_{-1}, \quad A_1 = a_1,$$
$$B_0 = a_0 - a_{-1}, \quad B_1 = a_1.$$

We have one constant less than before, but it will be seen that this is sufficient, for in fact A_1 and B_1 only differ by terms of an order which we are going to neglect. We assume a_1, a_{-1} to be small quantities.

Hence
$$x = (a_0 + a_{-1}) \cos \tau + a_1 \cos 3\tau,$$
$$y = (a_0 - a_{-1}) \sin \tau + a_1 \sin 3\tau.$$

Since $\quad \cos 3\tau = \quad 4 \cos^3 \tau - 3 \cos \tau = \quad \cos \tau \, (1 - 4 \sin^2 \tau),$

$$\sin 3\tau = -4 \sin^3 \tau + 3 \sin \tau = -\sin \tau \, (1 - 4 \cos^2 \tau),$$

we have
$$\left. \begin{array}{l} x = a_0 \cos \tau \left[1 + \dfrac{a_1 + a_{-1}}{a_0} - \dfrac{4a_1}{a_0} \sin^2 \tau \right] \\[3mm] y = a_0 \sin \tau \left[1 - \dfrac{a_1 + a_{-1}}{a_0} + \dfrac{4a_1}{a_0} \cos^2 \tau \right] \end{array} \right\} .$$

Neglecting powers of a_1, a_{-1} higher than the first, we deduce

$$r^2 = a_0^2 \left[1 + 2 \frac{a_1 + a_{-1}}{a_0} \cos 2\tau \right],$$

$$\frac{1}{r^3} = \frac{1}{a_0^3} \left[1 - 3 \frac{a_1 + a_{-1}}{a_0} \cos 2\tau \right]$$

$$= \frac{1}{a_0^3} \left[1 - 3 \frac{a_1 + a_{-1}}{a_0} + 6 \frac{a_1 + a_{-1}}{a_0} \sin^2 \tau \right]$$

$$= \frac{1}{a_0^3} \left[1 + 3 \frac{a_1 + a_{-1}}{a_0} - 6 \frac{a_1 + a_{-1}}{a_0} \cos^2 \tau \right];$$

$$\frac{\kappa x}{r^3} = \frac{\kappa}{a_0^2} \cos \tau \left[1 - \frac{2a_1 + 2a_{-1}}{a_0} + \frac{2a_1 + 6a_{-1}}{a_0} \sin^2 \tau \right],$$

$$\frac{\kappa y}{r^3} = \frac{\kappa}{a_0^2} \sin \tau \left[1 + \frac{2a_1 + 2a_{-1}}{a_0} - \frac{2a_1 + 6a_{-1}}{a_0} \cos^2 \tau \right],$$

$$\frac{d^2 x}{d\tau^2} = -\left[(a_0 + a_{-1}) \cos \tau + 9a_1 \cos 3\tau \right] = -\cos \tau \left[a_0 + 9a_1 + a_{-1} - 36a_1 \sin^2 \tau \right],$$

$$\frac{d^2 y}{d\tau^2} = -\left[(a_0 - a_{-1}) \sin \tau + 9a_1 \sin 3\tau \right] = -\sin \tau \left[a_0 - 9a_1 - a_{-1} + 36a_1 \cos^2 \tau \right].$$

With the required accuracy

$$-2m \frac{dy}{d\tau} = -2m a_0 \cos \tau, \quad 2m \frac{dx}{d\tau} = -2m a_0 \sin \tau, \quad \text{and} \quad 3m^2 x = 3m^2 a_0 \cos \tau.$$

Substituting these results in the differential equations, (6), we get

$$a_0 \cos \tau \left[-1 - \frac{9a_1 + a_{-1}}{a_0} + \frac{36a_1}{a_0} \sin^2 \tau - 2m \right.$$

$$\left. + \frac{\kappa}{a_0^3} \left(1 - \frac{2a_1 + 2a_{-1}}{a_0} + \frac{2a_1 + 6a_{-1}}{a_0} \sin^2 \tau \right) - 3m^2 \right] = 0,$$

$$a_0 \sin \tau \left[-1 + \frac{9a_1 + a_{-1}}{a_0} - \frac{36a_1}{a_0} \cos^2 \tau - 2m \right.$$

$$\left. + \frac{\kappa}{a_0^3} \left(1 + \frac{2a_1 + 2a_{-1}}{a_0} - \frac{2a_1 + 6a_{-1}}{a_0} \cos^2 \tau \right) \right] = 0.$$

Equating to zero the coefficients of $\cos \tau$, $\cos \tau \sin^2 \tau$, $\sin \tau$, $\sin \tau \cos^2 \tau$, we get

$$\left.\begin{aligned}
-1 - \frac{9a_1 + a_{-1}}{a_0} - 2m + \frac{\kappa}{a_0{}^3}\left(1 - \frac{2a_1 + 2a_{-1}}{a_0}\right) - 3m^2 &= 0 \\
-1 + \frac{9a_1 + a_{-1}}{a_0} - 2m + \frac{\kappa}{a_0{}^3}\left(1 + \frac{2a_1 + 2a_{-1}}{a_0}\right) &= 0 \\
\frac{36a_1}{a_0} + \frac{\kappa}{a_0{}^3}\left(\frac{2a_1 + 6a_{-1}}{a_0}\right) &= 0
\end{aligned}\right\}\dots\dots(7).$$

As there are only three equations for the determination of $\dfrac{\kappa}{a_0{}^3}$, $\dfrac{a_1}{a_0}$, $\dfrac{a_{-1}}{a_0}$ our assumption that $A_1 = B_1 = a_1$ is justified to the order of small quantities considered.

Half the sum and difference of the first two give

$$-1 - 2m - \tfrac{3}{2}m^2 + \frac{\kappa}{a_0{}^3} = 0,$$

$$\frac{9a_1 + a_{-1}}{a_0} + \frac{2\kappa}{a_0{}^3}\frac{a_1 + a_{-1}}{a_0} + \tfrac{3}{2}m^2 = 0.$$

Therefore

$$\frac{\kappa}{a_0{}^3} = 1 + 2m + \tfrac{3}{2}m^2,$$

$$\frac{11a_1}{a_0} + \frac{3a_{-1}}{a_0} = -\tfrac{3}{2}m^2, \text{ to our order of accuracy, viz. } m^2;$$

also

$$\frac{19a_1}{a_0} + \frac{3a_{-1}}{a_0} = 0, \text{ from the third equation};$$

$$\therefore \frac{8a_1}{a_0} = \tfrac{3}{2}m^2,$$

$$\left.\begin{aligned}
\frac{a_1}{a_0} &= \tfrac{3}{16}m^2, \quad \frac{a_{-1}}{a_0} = -\tfrac{19}{16}m^2 \\
\frac{\kappa}{a_0{}^3} &= 1 + 2m + \tfrac{3}{2}m^2
\end{aligned}\right\}\dots\dots\dots\dots\dots\dots(8).$$

Hence

$$x = a_0\left[(1 - \tfrac{19}{16}m^2)\cos \tau + \tfrac{3}{16}m^2 \cos 3\tau\right],$$

$$y = a_0\left[(1 + \tfrac{19}{16}m^2)\sin \tau + \tfrac{3}{16}m^2 \sin 3\tau\right],$$

or perhaps more conveniently for future work

$$\left.\begin{aligned}
x &= a_0 \cos \tau \left[1 - m^2 - \tfrac{3}{4}m^2 \sin^2 \tau\right] \\
y &= a_0 \sin \tau \left[1 + m^2 + \tfrac{3}{4}m^2 \cos^2 \tau\right]
\end{aligned}\right\}\dots\dots\dots\dots(9).$$

It will be seen that these are the equations to an oval curve, the semi-axes of which are $a_0(1 - m^2)$, $a_0(1 + m^2)$ along and perpendicular to the line joining the earth and sun. If r, θ be the polar coordinates of a point on the curve,

$$r^2 = a_0{}^2\left[1 - 2m^2 \cos 2\tau\right],$$

giving

$$r = a_0\left[1 - m^2 \cos 2\tau\right] \dots\dots\dots\dots\dots(10).$$

Also
$$\tan \theta = \frac{y}{x} = \tan \tau \left[1 + 2m^2 + \tfrac{3}{4}m^2\right]$$

$$= (1 + \tfrac{11}{4}\,m^2) \tan \tau.$$

$$\therefore \quad \tan(\theta - \tau) = \frac{\tan \tau}{1 + \tan^2 \tau} \cdot \tfrac{11}{4}m^2 = \tfrac{11}{8}m^2 \sin 2\tau,$$

giving
$$\theta = \tau + \tfrac{11}{8}m^2 \sin 2\tau \quad\dots\dots\dots\dots\dots\dots(11).$$

If a be the mean distance corresponding to a mean motion n in an undisturbed orbit, Kepler's third law gives

$$n^2 a^3 = E + M = \kappa \nu^2 \quad\dots\dots\dots\dots\dots\dots\dots(12).$$

But
$$\frac{n}{\nu} = \frac{n - n' + n'}{n - n'} = 1 + m.$$

Hence
$$(1 + m)^2 a^3 = \kappa = a_0{}^3 (1 + 2m + \tfrac{3}{2}m^2),$$

$$\frac{a_0{}^3}{a^3} = \frac{1 + 2m + m^2}{1 + 2m + m^2 + \tfrac{1}{2}m^2},$$

and
$$a_0 = a(1 - \tfrac{1}{6}m^2) \quad\dots\dots\dots\dots\dots\dots\dots(13).$$

This is a relation between a_0 and the undisturbed mean distance.

§4. Differential Equations
for Small Displacements from the Variational Curve.

If the solar perturbations were to vanish, m would be zero and we should have $x = a_0 \cos \tau,\ y = a_0 \sin \tau$ so that the orbit would be a circle. We may therefore consider the orbit already found as a circular orbit distorted by solar influence. [We have indeed put $\Omega = 0$, but the terms neglected are small and need not be considered at present.] As the circular orbit is only a special solution of the problem of two bodies, we should not expect the variational curve to give the actual motion of the moon. In fact it is known that the moon moves rather in an ellipse of eccentricity $\frac{1}{20}$ than in a circle or variational curve. The latter therefore will only serve as an approximation to the real orbit in the same way as a circle serves as an approximation to an ellipse. An ellipse of small eccentricity can be obtained by "free oscillations" about a circle, and what we proceed to do is to determine free oscillations about the variational curve. We thus introduce two new arbitrary constants —determining the amplitude and phase of the oscillations—and so get the general solution of our differential equations (6). The procedure is exactly similar to that used in dynamics for the discussion of small oscillations about a steady state, i.e., the moon is initially supposed to lie near the variational curve, and its subsequent motion is determined relatively to this curve. At first only first powers of the small quantities will be used—an approximation which corresponds to the first powers of the eccentricity in the elliptic theory. If required, further approximations can be made.

Suppose then that x, y are the coordinates of a point on the variational curve which we have found to satisfy the differential equations of motion and that $x + \delta x$, $y + \delta y$ are the coordinates of the moon in her actual orbit, then since x, y satisfy the equations it is clear that the equations to be satisfied by δx, δy. are

$$\left. \begin{aligned} \frac{d^2}{d\tau^2}\,\delta x - 2\mathrm{m}\,\frac{d}{d\tau}\,\delta y + \kappa\delta\left(\frac{x}{r^3}\right) - 3\mathrm{m}^2\delta x &= 0 \\[2mm] \frac{d^2}{d\tau^2}\,\delta y + 2\mathrm{m}\,\frac{d}{d\tau}\,\delta x + \kappa\delta\left(\frac{y}{r^3}\right) \quad\quad &= 0 \end{aligned} \right\} \quad\dots\dots\dots(14).$$

Now it is not convenient to proceed immediately from these equations as you may see by considering how you would proceed if the orbit of reference were a simple undisturbed circle. The obvious course is to replace δx, δy by normal and tangential displacements δp, δs.

Suppose then that ϕ denotes the inclination of the outward normal of the variational curve to the x-axis. Then we have

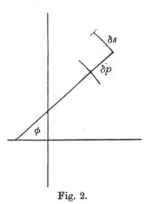

Fig. 2.

$$\left. \begin{aligned} \delta x &= \delta p \cos\phi - \delta s \sin\phi \\ \delta y &= \delta p \sin\phi + \delta s \cos\phi \end{aligned} \right\} \quad\dots\dots(15).$$

Multiply the first differential equation (14) by $\cos\phi$ and the second by $\sin\phi$ and add; and again multiply the first by $\sin\phi$ and the second by $\cos\phi$ and subtract. We have

$$\left. \begin{aligned} &\cos\phi\,\frac{d^2\,\delta x}{d\tau^2} + \sin\phi\,\frac{d^2\delta y}{d\tau^2} - 2\mathrm{m}\left[\cos\phi\,\frac{d\delta y}{d\tau} - \sin\phi\,\frac{d\delta x}{d\tau}\right] \\ &\quad + \kappa\cos\phi\delta\left(\frac{x}{r^3}\right) + \kappa\sin\phi\delta\left(\frac{y}{r^3}\right) - 3\mathrm{m}^2\cos\phi\delta x = 0 \\[3mm] &-\sin\phi\,\frac{d^2\,\delta x}{d\tau^2} + \cos\phi\,\frac{d^2\delta y}{d\tau^2} + 2\mathrm{m}\left[\sin\phi\,\frac{d\delta y}{d\tau} + \cos\phi\,\frac{d\delta x}{d\tau}\right] \\ &\quad - \kappa\sin\phi\delta\left(\frac{x}{r^3}\right) + \kappa\cos\phi\delta\left(\frac{y}{r^3}\right) + 3\mathrm{m}^2\sin\phi\delta x = 0 \end{aligned} \right\} \dots(16).$$

Now we have from (15)

$$\delta p = \delta x \cos\phi + \delta y \sin\phi, \quad \delta s = \delta x \sin\psi + \delta y \cos\phi.$$

Therefore

$$\frac{d\,\delta p}{d\tau} = \quad\cos\phi\,\frac{d\,\delta x}{d\tau} + \sin\phi\,\frac{d\,\delta y}{d\tau} + (-\,\delta x \sin\phi + \delta y \cos\phi)\frac{d\phi}{d\tau},$$

$$\frac{d\,\delta s}{d\tau} = -\sin\phi\,\frac{d\,\delta x}{d\tau} + \cos\phi\,\frac{d\,\delta y}{d\tau} - (\quad\delta x \cos\phi + \delta y \sin\phi)\frac{d\phi}{d\tau}.$$

Hence the two expressions which occur in the second group of terms of (16) are

$$\cos\phi\,\frac{d\,\delta y}{d\tau} - \sin\phi\,\frac{d\,\delta x}{d\tau} = \frac{d\,\delta s}{d\tau} + \delta p\,\frac{d\phi}{d\tau},$$

$$\sin\phi\,\frac{d\,\delta y}{d\tau} + \cos\phi\,\frac{d\,\delta x}{d\tau} = \frac{d\,\delta p}{d\tau} - \delta s\,\frac{d\phi}{d\tau}.$$

When we differentiate these again, we obtain the first group of terms in (16). Inverting the order of the equations we have

$$\cos\phi\,\frac{d^2\,\delta x}{d\tau^2} + \sin\phi\,\frac{d^2\,\delta y}{d\tau^2}$$

$$= \frac{d^2\,\delta p}{d\tau^2} - \frac{d\,\delta s}{d\tau}\frac{d\phi}{d\tau} - \delta s\,\frac{d^2\phi}{d\tau^2} - \left(\cos\phi\,\frac{d\,\delta y}{d\tau} - \sin\phi\,\frac{d\,\delta x}{d\tau}\right)\frac{d\phi}{d\tau}$$

$$= \frac{d^2\,\delta p}{d\tau^2} - 2\,\frac{d\,\delta s}{d\tau}\frac{d\phi}{d\tau} - \delta p\left(\frac{d\phi}{d\tau}\right)^2 - \delta s\,\frac{d^2\phi}{d\tau^2},$$

$$-\sin\phi\,\frac{d^2\,\delta x}{d\tau^2} + \cos\phi\,\frac{d^2\,\delta y}{d\tau^2}$$

$$= \frac{d^2\,\delta s}{d\tau^2} + \frac{d\,\delta p}{d\tau}\frac{d\phi}{d\tau} + \delta p\,\frac{d^2\phi}{d\tau^2} + \left(\sin\phi\,\frac{d\,\delta y}{d\tau} + \cos\phi\,\frac{d\,\delta x}{d\tau}\right)\frac{d\phi}{d\tau}$$

$$= \frac{d^2\,\delta s}{d\tau^2} + 2\,\frac{d\,\delta p}{d\tau}\frac{d\phi}{d\tau} - \delta s\left(\frac{d\phi}{d\tau}\right)^2 + \delta p\,\frac{d^2\phi}{d\tau^2}.$$

Substituting in (16), we have as our equations

$$\frac{d^2\,\delta p}{d\tau^2} - \delta p\left[\left(\frac{d\phi}{d\tau}\right)^2 + 2\mathrm{m}\,\frac{d\phi}{d\tau}\right] - 2\,\frac{d\,\delta s}{d\tau}\left(\frac{d\phi}{d\tau} + \mathrm{m}\right) - \delta s\,\frac{d^2\phi}{d\tau^2}$$

$$+ \kappa\cos\phi\,\delta\left(\frac{x}{r^3}\right) + \kappa\sin\phi\,\delta\left(\frac{y}{r^3}\right) - 3\mathrm{m}^2\cos\phi\,\delta x = 0$$

$$\left.\begin{array}{l}\end{array}\right\}\ldots(17).$$

$$\frac{d^2\,\delta s}{d\tau^2} - \delta s\left[\left(\frac{d\phi}{d\tau}\right)^2 + 2\mathrm{m}\,\frac{d\phi}{d\tau}\right] + 2\,\frac{d\,\delta p}{d\tau}\left(\frac{d\phi}{d\tau} + \mathrm{m}\right) + \delta p\,\frac{d^2\phi}{d\tau^2}$$

$$- \kappa\sin\phi\,\delta\left(\frac{x}{r^3}\right) + \kappa\cos\phi\,\delta\left(\frac{y}{r^3}\right) + 3\mathrm{m}^2\sin\phi\,\delta x = 0$$

Variation of the Jacobian integral

$$V^2 = \left(\frac{dx}{d\tau}\right)^2 + \left(\frac{dy}{d\tau}\right)^2 = \frac{2\kappa}{r} + 3\mathrm{m}^2 x^2 + C$$

gives

$$\frac{dx}{d\tau}\frac{d\,\delta x}{d\tau} + \frac{dy}{d\tau}\frac{d\,\delta y}{d\tau} = -\frac{\kappa}{r^2}\,\delta r + 3\mathrm{m}^2 x\,\delta x\,*.$$

Now

$$\frac{dx}{d\tau} = -V\sin\phi,\qquad \frac{dy}{d\tau} = V\cos\phi,$$

---

* We could introduce a term δC, but the variation of the orbit which we are introducing is one for which C is unaltered.

and
$$\frac{d\,\delta x}{d\tau} = \cos\phi\,\frac{d\,\delta p}{d\tau} - \delta s\cos\phi\,\frac{d\phi}{d\tau} - \sin\phi\,\frac{d\,\delta s}{d\tau} - \sin\phi\,\delta p\,\frac{d\phi}{d\tau},$$

$$\frac{d\,\delta y}{d\tau} = \sin\phi\,\frac{d\,\delta p}{d\tau} - \delta s\sin\phi\,\frac{d\phi}{d\tau} + \cos\phi\,\frac{d\,\delta s}{d\tau} + \cos\phi\,\delta p\,\frac{d\phi}{d\tau}.$$

Hence
$$\frac{dx}{d\tau}\frac{d\,\delta x}{d\tau} + \frac{dy}{d\tau}\frac{d\,\delta y}{d\tau} = V\left(\frac{d\,\delta s}{d\tau} + \delta p\,\frac{d\phi}{d\tau}\right).$$

Also

$$-\frac{\kappa\,\delta r}{r^2} = -\frac{\kappa}{r^3}(x\,\delta x + y\,\delta y) = -\frac{\kappa x}{r^3}(\delta p\cos\phi - \delta s\sin\phi) - \frac{\kappa y}{r^3}(\delta p\sin\phi + \delta s\cos\phi)$$

$$= -\frac{\kappa}{r^3}[\delta p\,(x\cos\phi + y\sin\phi) + \delta s\,(-x\sin\phi + y\cos\phi)].$$

Thus, retaining the term $3m^2 x\,\delta x$ in its original form, the varied Jacobian integral becomes

$$V\left(\frac{d\,\delta s}{d\tau} + \delta p\,\frac{d\phi}{d\tau}\right)$$

$$= -\frac{\kappa}{r^3}[\delta p\,(x\cos\phi + y\sin\phi) + \delta s\,(-x\sin\phi + y\cos\phi)] + 3m^2 x\,\delta x \ldots(18).$$

Before we can solve the differential equations (17)·for δp, δs we require to express all the other variables occurring in them in terms of τ by means of the equations obtained in § 3.

§ 5. *Transformation of the equations in* § 4.

We desire to transform the differential equations (17) so that the only variables involved will be δp, δs, τ. We shall then be in a position to solve for δp, δs in terms of τ.

We have
$$r\,\delta r = x\,\delta x + y\,\delta y - (x\cos\phi + y\sin\phi)\,\delta p + (-x\sin\phi + y\cos\phi)\,\delta s.$$

Hence

$$\cos\phi\,\delta\left(\frac{x}{r^3}\right) + \sin\phi\,\delta\left(\frac{y}{r^3}\right)$$

$$= \frac{1}{r^3}(\delta x\cos\phi + \delta y\sin\phi) - \frac{3}{r^5}(x\cos\phi + y\sin\phi)\,r\,\delta r$$

$$= \frac{\delta p}{r^3} - \frac{3}{r^5}\left[(x^2\cos^2\phi + y^2\sin^2\phi + 2xy\sin\phi\cos\phi)\,\delta p\right.$$

$$\left. + (-x^2\sin\phi\cos\phi + xy\cos^2\phi - xy\sin^2\phi + y^2\sin\phi\cos\phi)\,\delta s\right]$$

$$= \frac{\delta p}{r^3} - \frac{3}{r^5}\left[\{\tfrac{1}{2}(x^2 + y^2) + \tfrac{1}{2}(x^2 - y^2)\cos 2\phi + xy\sin 2\phi\}\,\delta p\right.$$

$$\left. + \{-\tfrac{1}{2}(x^2 - y^2)\sin 2\phi + xy\cos 2\phi\}\,\delta s\right]$$

$$= \frac{\delta p}{r^3}\left[-\tfrac{1}{2} - \tfrac{3}{2}\frac{x^2 - y^2}{r^2}\cos 2\phi - \frac{3xy}{r^2}\sin 2\phi\right]$$

$$- \frac{3\,\delta s}{r^3}\left[\frac{xy}{r^2}\cos 2\phi - \tfrac{1}{2}\frac{x^2 - y^2}{r^2}\sin 2\phi\right] \ldots(19),$$

$$-\sin\phi\delta\left(\frac{x}{r^3}\right)+\cos\phi\delta\left(\frac{y}{r^3}\right)$$

$$=\frac{1}{r^3}\left(-\delta x\sin\phi+\delta y\cos\phi\right)-\frac{3}{r^5}\left(-x\sin\phi+y\cos\phi\right)r\,\delta r$$

$$=\frac{\delta s}{r^3}-\frac{3}{r^5}\left[\left(-x^2\sin\phi\cos\phi-xy\sin^2\phi+xy\cos^2\phi+y^2\sin\phi\cos\phi\right)\delta p\right.$$

$$\left.+\left(x^2\sin^2\phi+y^2\cos^2\phi-2xy\sin\phi\cos\phi\right)\delta s\right]$$

$$=\frac{\delta s}{r^3}-\frac{3}{r^5}\left[\left\{-\tfrac{1}{2}\left(x^2-y^2\right)\sin 2\phi+xy\cos 2\phi\right\}\delta p\right.$$

$$\left.+\left\{\tfrac{1}{2}\left(x^2+y^2\right)-\tfrac{1}{2}\left(x^2-y^2\right)\cos 2\phi-xy\sin 2\phi\right\}\delta s\right]$$

$$=-\frac{3\delta p}{r^3}\left[\frac{xy}{r^2}\cos 2\phi-\tfrac{1}{2}\frac{x^2-y^2}{r^2}\sin 2\phi\right]$$

$$+\frac{\delta s}{r^3}\left[-\tfrac{1}{2}+\tfrac{3}{2}\frac{x^2-y^2}{r^2}\cos 2\phi+\frac{3xy}{r^2}\sin 2\phi\right]\ldots(20).$$

We shall consider the terms $3m^2\delta x\genfrac{}{}{0pt}{}{\cos}{\sin}\phi$ later (p. 33).

The next step is to substitute throughout the differential equations (17) the values of x, y and ϕ which correspond to the undisturbed orbit. For simplicity in writing we drop the linear factor a_0. It can be easily introduced when required.

We have already found, in (9),

$$x=\cos\tau\left(1-\tfrac{19}{16}m^2\right)+\tfrac{3}{16}m^2\cos 3\tau=\cos\tau\left(1-m^2-\tfrac{3}{4}m^2\sin^2\tau\right),$$

$$y=\sin\tau\left(1+\tfrac{19}{16}m^2\right)+\tfrac{3}{16}m^2\sin 3\tau=\sin\tau\left(1+m^2+\tfrac{3}{4}m^2\cos^2\tau\right).$$

Then

$$\frac{dx}{d\tau}=-\sin\tau\left(1-\tfrac{7}{4}m^2+\tfrac{9}{4}m^2\cos^2\tau\right)=-\sin\tau\left(1+\tfrac{1}{2}m^2-\tfrac{9}{4}m^2\sin^2\tau\right),$$

$$\frac{dy}{d\tau}=\cos\tau\left(1+\tfrac{7}{4}m^2-\tfrac{9}{4}m^2\sin^2\tau\right)=\cos\tau\left(1-\tfrac{1}{2}m^2+\tfrac{9}{4}m^2\cos^2\tau\right).$$

Whence

$$V^2=\left(\frac{dx}{d\tau}\right)^2+\left(\frac{dy}{d\tau}\right)^2=\sin^2\tau\left(1+m^2-\tfrac{9}{2}m^2\sin^2\tau\right)+\cos^2\tau\left(1-m^2+\tfrac{9}{2}m^2\cos^2\tau\right)$$

$$=1-m^2\cos 2\tau+\tfrac{9}{2}m^2\cos 2\tau=1+\tfrac{7}{2}m^2\cos 2\tau$$

$$=1+\tfrac{7}{2}m^2-7m^2\sin^2\tau=1-\tfrac{7}{2}m^2+7m^2\cos^2\tau.$$

Therefore

$$\frac{1}{V}=1+\tfrac{7}{4}m^2-\tfrac{7}{2}m^2\cos^2\tau=1-\tfrac{7}{4}m^2+\tfrac{7}{2}m^2\sin^2\tau=1-\tfrac{7}{4}m^2\cos 2\tau.$$

Now $$\sin \phi = -\frac{1}{V}\frac{dx}{d\tau}, \quad \cos \phi = \frac{1}{V}\frac{dy}{d\tau}.$$

Therefore

$$\sin \phi = \sin \tau \left(1 + \tfrac{1}{2}m^2 - \tfrac{9}{4}m^2 \sin^2 \tau - \tfrac{7}{4}m^2 + \tfrac{7}{2}m^2 \sin^2 \tau\right)$$

$$= \sin \tau \left(1 - \tfrac{5}{4}m^2 + \tfrac{5}{4}m^2 \sin^2 \tau\right) = \sin \tau \left(1 - \tfrac{5}{4}m^2 \cos^2 \tau\right),$$

$$\cos \phi = \cos \tau \left(1 - \tfrac{1}{2}m^2 + \tfrac{9}{4}m^2 \cos^2 \tau + \tfrac{7}{4}m^2 - \tfrac{7}{2}m^2 \cos^2 \tau\right)$$

$$= \cos \tau \left(1 + \tfrac{5}{4}m^2 - \tfrac{5}{4}m^2 \cos^2 \tau\right) = \cos \tau \left(1 + \tfrac{5}{4}m^2 \sin^2 \tau\right);$$

$$\sin 2\phi = \sin 2\tau \left(1 - \tfrac{5}{4}m^2 \cos 2\tau\right),$$

$$\cos 2\phi = \cos 2\tau + \tfrac{5}{4}m^2 \sin^2 2\tau ;$$

$$\cos \phi \frac{d\phi}{d\tau} = \cos \tau \left(1 - \tfrac{5}{4}m^2 + \tfrac{15}{4}m^2 \sin^2 \tau\right),$$

$$\sin \phi \frac{d\phi}{d\tau} = -\sin \tau \left(1 + \tfrac{5}{4}m^2 - \tfrac{15}{4}m^2 \cos^2 \tau\right).$$

Summing the squares of these,

$$\left(\frac{d\phi}{d\tau}\right)^2 = \cos^2 \tau \left(1 - \tfrac{5}{2}m^2 + \tfrac{15}{2}m^2 \sin^2 \tau\right) + \sin^2 \tau \left(1 + \tfrac{5}{2}m^2 - \tfrac{15}{2}m^2 \cos^2 \tau\right)$$

$$= 1 - \tfrac{5}{2}m^2 \cos 2\tau,$$

and thence $$\frac{d\phi}{d\tau} = 1 - \tfrac{5}{4}m^2 \cos 2\tau \dots\dots\dots\dots\dots\dots(21).$$

Differentiating again $$\frac{d^2\phi}{d\tau^2} = \tfrac{5}{2}m^2 \sin 2\tau.$$

We are now in a position to evaluate all the earlier terms in the differential equations (17).

Thus

$$\frac{d^2 \delta p}{d\tau^2} - \delta p \left[\left(\frac{d\phi}{d\tau}\right)^2 + 2m\frac{d\phi}{d\tau}\right] - 2\frac{d\delta s}{d\tau}\left(\frac{d\phi}{d\tau}+m\right) - \delta s \frac{d^2\phi}{d\tau^2}$$

$$= \frac{d^2 \delta p}{d\tau^2} + \delta p \left[-1 + \tfrac{5}{2}m^2 \cos 2\tau - 2m\right] - 2\frac{d\delta s}{d\tau}\left(1 + m - \tfrac{5}{4}m^2 \cos 2\tau\right)$$

$$- \tfrac{5}{2}m^2 \sin 2\tau\, \delta s$$

$$\frac{d^2 \delta s}{d\tau^2} - \delta s \left[\left(\frac{d\phi}{d\tau}\right)^2 + 2m\frac{d\phi}{d\tau}\right] + 2\frac{d\delta p}{d\tau}\left(\frac{d\phi}{d\tau}+m\right) + \delta p \frac{d^2\phi}{d\tau^2}$$

$$= \frac{d^2 \delta s}{d\tau^2} + \delta s \left[-1 + \tfrac{5}{2}m^2 \cos 2\tau - 2m\right] + 2\frac{d\delta p}{d\tau}\left(1 + m - \tfrac{5}{4}m^2 \cos 2\tau\right)$$

$$+ \tfrac{5}{2}m^2 \sin 2\tau\, \delta p$$

$$\left.\right\}\dots(22).$$

We now have to evaluate the several terms involving x and y in (18), (19), (20).

$$x \cos \phi + y \sin \phi = \cos^2 \tau \left(1 - m^2 - \tfrac{3}{4}m^2 \sin^2 \tau + \tfrac{5}{4}m^2 \sin^2 \tau\right)$$
$$+ \sin^2 \tau \left(1 + m^2 + \tfrac{3}{4}m^2 \cos^2 \tau - \tfrac{5}{4}m^2 \cos^2 \tau\right)$$
$$= 1 - m^2 \cos 2\tau,$$

$$-x \sin \phi + y \cos \phi = -\sin \tau \cos \tau \left(1 - m^2 - \tfrac{3}{4}m^2 \sin^2 \tau - \tfrac{5}{4}m^2 \cos^2 \tau\right)$$
$$+ \sin \tau \cos \tau \left(1 + m^2 + \tfrac{3}{4}m^2 \cos^2 \tau + \tfrac{5}{4}m^2 \sin^2 \tau\right)$$
$$= 2m^2 \sin 2\tau ;$$

$$r^2 = x^2 + y^2 = 1 - 2m^2 \cos 2\tau,$$

$$x^2 - y^2 = \cos^2 \tau \left(1 - 2m^2 - \tfrac{3}{2}m^2 \sin^2 \tau\right)$$
$$- \sin^2 \tau \left(1 + 2m^2 + \tfrac{3}{2}m^2 \cos^2 \tau\right)$$
$$= \cos 2\tau - 2m^2 - \tfrac{3}{4}m^2 \sin^2 2\tau,$$

$$xy = \tfrac{1}{2} \sin 2\tau \left(1 + \tfrac{3}{4}m^2 \cos 2\tau\right) ;$$

$$(x^2 - y^2) \cos 2\phi = \cos^2 2\tau - 2m^2 \cos 2\tau - \tfrac{3}{4}m^2 \sin^2 2\tau \cos 2\tau$$
$$+ \tfrac{5}{4}m^2 \sin^2 2\tau \cos 2\tau$$
$$= \cos 2\tau \left(\cos 2\tau - 2m^2 + \tfrac{1}{2}m^2 \sin^2 2\tau\right),$$

$$(x^2 - y^2) \sin 2\phi = \sin 2\tau \left(\cos 2\tau - 2m^2 - \tfrac{3}{4}m^2 \sin^2 2\tau - \tfrac{5}{4}m^2 \cos^2 2\tau\right)$$
$$= \sin 2\tau \left(\cos 2\tau - \tfrac{11}{4}m^2 - \tfrac{1}{2}m^2 \cos^2 2\tau\right) ;$$

$$xy \cos 2\phi = \tfrac{1}{2} \sin 2\tau \left(\cos 2\tau + \tfrac{5}{4}m^2 \sin^2 2\tau + \tfrac{3}{4}m^2 \cos^2 2\tau\right)$$
$$= \tfrac{1}{2} \sin 2\tau \left(\cos 2\tau + \tfrac{5}{4}m^2 - \tfrac{1}{2}m^2 \cos^2 2\tau\right),$$

$$xy \sin 2\phi = \tfrac{1}{2} \sin^2 2\tau \left(1 - \tfrac{5}{4}m^2 \cos 2\tau + \tfrac{3}{4}m^2 \cos 2\tau\right)$$
$$= \tfrac{1}{2} \sin^2 2\tau \left(1 - \tfrac{1}{2}m^2 \cos 2\tau\right).$$

Therefore

$$\tfrac{1}{2}(x^2 - y^2) \cos 2\phi + xy \sin 2\phi$$
$$= \tfrac{1}{2}\cos^2 2\tau - m^2 \cos 2\tau + \tfrac{1}{4}m^2 \sin^2 2\tau \cos 2\tau + \tfrac{1}{2}\sin^2 2\tau - \tfrac{1}{4}m^2 \sin^2 2\tau \cos 2\tau$$
$$= \tfrac{1}{2}(1 - 2m^2 \cos 2\tau) = \tfrac{1}{2}r^2,$$

$$\therefore \quad -\tfrac{1}{2} \mp \tfrac{3}{2}\frac{x^2 - y^2}{r^2} \cos 2\phi \mp \frac{3xy}{r^2} \sin 2\phi = -\tfrac{1}{2} \mp \tfrac{3}{2} = -2 \text{ or } +1.$$

These are the coefficients of $\dfrac{\delta p}{r^3}$ in the expression (19) for

$$\cos \phi \delta \left(\frac{x}{r^3}\right) + \sin \phi \delta \left(\frac{y}{r^3}\right),$$

and of $\dfrac{\delta s}{r^3}$ in the expression (20) for $-\sin \phi \delta \left(\dfrac{x}{r^3}\right) + \cos \phi \delta \left(\dfrac{y}{r^3}\right).$

Again

$$-\tfrac{1}{2}(x^2 - y^2)\sin 2\phi + xy\cos 2\phi$$
$$= -\tfrac{1}{2}\sin 2\tau\,(\cos 2\tau - \tfrac{11}{4}m^2 - \tfrac{1}{2}m^2\cos^2 2\tau)$$
$$+ \tfrac{1}{2}\sin 2\tau\,(\cos 2\tau + \tfrac{5}{4}m^2 - \tfrac{1}{2}m^2\cos^2 2\tau)$$
$$= 2m^2\sin 2\tau.$$

Then since to the order zero, $r^2 = 1$, we have

$$3\left(\frac{xy}{r^2}\cos 2\phi - \tfrac{1}{2}\frac{x^2-y^2}{r^2}\sin 2\phi\right) = 6m^2\sin 2\tau.$$

This is the coefficient of $-\dfrac{\delta s}{r^3}$ in $\cos\phi\,\delta\left(\dfrac{x}{r^3}\right) + \sin\phi\,\delta\left(\dfrac{y}{r^3}\right)$ and of $-\dfrac{\delta p}{r^3}$ in

$$-\sin\phi\,\delta\left(\frac{x}{r^3}\right) + \cos\phi\,\delta\left(\frac{y}{r^3}\right).$$

Hence we have

$$\left.\begin{aligned}
\cos\phi\,\delta\left(\frac{x}{r^3}\right) + \sin\phi\,\delta\left(\frac{y}{r^3}\right) &= -2\frac{\delta p}{r^3} - \frac{6m^2}{r^3}\delta s\sin 2\tau\\
&= -2\delta p\,(1 + 3m^2\cos 2\tau) - 6m^2\delta s\sin 2\tau\\
-\sin\phi\,\delta\left(\frac{x}{r^3}\right) + \cos\phi\,\delta\left(\frac{y}{r^3}\right) &= -\frac{\delta p}{r^3}.6m^2\sin 2\tau + \frac{\delta s}{r^3}\\
&= -6m^2\,\delta p\sin 2\tau + \delta s\,(1 + 3m^2\cos 2\tau)
\end{aligned}\right\}\dots(23).$$

These two expressions are to be multiplied by κ in the differential equations (17).

The other terms which occur in the differential equations are $-3m^2\cos\phi\,\delta x$ and $+3m^2\sin\phi\,\delta x$.

Since m^2 occurs in the coefficient we need only go to the order zero of small quantities in $\cos\phi\,\delta x$ and $\sin\phi\,\delta x$.

Thus

$$3m^2\,\delta x\cos\phi = 3m^2\,(\delta p\cos\tau - \delta s\sin\tau)\cos\tau = \tfrac{3}{2}m^2\,\delta p\,(1+\cos 2\tau) - \tfrac{3}{2}m^2\delta s\sin 2\tau,$$
$$3m^2\,\delta x\sin\phi = 3m^2\,(\delta p\cos\tau - \delta s\sin\tau)\sin\tau = \tfrac{3}{2}m^2\,\delta p\sin 2\tau - \tfrac{3}{2}m^2\,\delta s\,(1-\cos 2\tau).$$

Now $\kappa = 1 + 2m + \tfrac{3}{2}m^2$, and hence

$$\kappa\cos\phi\,\delta\left(\frac{x}{r^3}\right) + \kappa\sin\phi\,\delta\left(\frac{y}{r^3}\right) - 3m^2\delta x\cos\phi$$
$$= -2\delta p\,(1 + 3m^2\cos 2\tau + 2m + \tfrac{3}{2}m^2) - 6m^2\delta s\sin 2\tau$$
$$- \tfrac{3}{2}m^2\,\delta p\,(1+\cos 2\tau) + \tfrac{3}{2}m^2\delta s\sin 2\tau$$
$$= -2\delta p\,[1 + 2m + \tfrac{9}{4}m^2 + \tfrac{15}{4}m^2\cos 2\tau] - \tfrac{9}{2}m^2\delta s\sin 2\tau,$$

$$-\kappa \sin \phi \delta \left(\frac{x}{r^3}\right) + \kappa \cos \phi \delta \left(\frac{y}{r^3}\right) + 3\mathrm{m}^2 \delta x \sin \phi$$

$$= -6\mathrm{m}^2 \delta p \sin 2\tau + \delta s \left(1 + 2\mathrm{m} + \tfrac{3}{2}\mathrm{m}^2 + 3\mathrm{m}^2 \cos 2\tau\right)$$

$$+ \tfrac{3}{2}\mathrm{m}^2 \delta p \sin 2\tau - \delta s \left(\tfrac{3}{2}\mathrm{m}^2 - \tfrac{3}{2}\mathrm{m}^2 \cos 2\tau\right)$$

$$= -\tfrac{9}{2}\mathrm{m}^2 \delta p \sin 2\tau + \delta s \left(1 + 2\mathrm{m} + \tfrac{9}{2}\mathrm{m}^2 \cos 2\tau\right).$$

Hence

$$\frac{d^2 \delta p}{d\tau^2} - \delta p \left[\left(\frac{d\phi}{d\tau}\right)^2 + 2\mathrm{m}\frac{d\phi}{d\tau}\right] - 2\frac{d\delta s}{d\tau}\left(\frac{d\phi}{d\tau} + \mathrm{m}\right) - \delta s \frac{d^2 \phi}{d\tau^2}$$

$$+ \kappa \cos \phi \delta \left(\frac{x}{r^3}\right) + \kappa \sin \phi \delta \left(\frac{y}{r^3}\right) - 3\mathrm{m}^2 \cos \phi \delta x = 0$$

becomes

$$\frac{d^2 \delta p}{d\tau^2} - \delta p \left[1 + 2\mathrm{m} - \tfrac{5}{2}\mathrm{m}^2 \cos 2\tau\right] - 2\frac{d\delta s}{d\tau}\left(1 + \mathrm{m} - \tfrac{5}{4}\mathrm{m}^2 \cos 2\tau\right) - \tfrac{5}{2}\mathrm{m}^2 \delta s \sin 2\tau$$

$$- 2\delta p \left[1 + 2\mathrm{m} + \tfrac{9}{4}\mathrm{m}^2 + \tfrac{15}{4}\mathrm{m}^2 \cos 2\tau\right] - \tfrac{9}{2}\mathrm{m}^2 \delta s \sin 2\tau = 0$$

or $\quad \dfrac{d^2 \delta p}{d\tau^2} - \delta p \left[3 + 6\mathrm{m} + \tfrac{9}{2}\mathrm{m}^2 + 5\mathrm{m}^2 \cos 2\tau\right] - 2\dfrac{d\delta s}{d\tau}\left(1 + \mathrm{m} - \tfrac{5}{4}\mathrm{m}^2 \cos 2\tau\right)$

$$- 7\mathrm{m}^2 \delta s \sin 2\tau = 0 \ldots (24).$$

This is the first of our equations transformed.

Again the second equation is

$$\frac{d^2 \delta s}{d\tau^2} - \delta s \left[\left(\frac{d\phi}{d\tau}\right)^2 + 2\mathrm{m}\frac{d\phi}{d\tau}\right] + 2\frac{d\delta p}{d\tau}\left(\frac{d\phi}{d\tau} + \mathrm{m}\right) + \delta p \frac{d^2 \phi}{d\tau^2}$$

$$- \kappa \sin \phi \delta \left(\frac{x}{r^3}\right) + \kappa \cos \phi \delta \left(\frac{y}{r^3}\right) + 3\mathrm{m}^2 \sin \phi \delta x = 0,$$

and it becomes

$$\frac{d^2 \delta s}{d\tau^2} + \delta s \left(-1 - 2\mathrm{m} + \tfrac{5}{2}\mathrm{m}^2 \cos 2\tau\right) + 2\frac{d\delta p}{d\tau}\left(1 + \mathrm{m} - \tfrac{5}{4}\mathrm{m}^2 \cos 2\tau\right) + \tfrac{5}{2}\mathrm{m}^2 \,\delta p \sin 2\tau$$

$$- \tfrac{9}{2}\mathrm{m}^2 \delta p \sin 2\tau + \delta s \left(1 + 2\mathrm{m} + \tfrac{9}{2}\mathrm{m}^2 \cos 2\tau\right) = 0.$$

Whence

$$\frac{d^2 \delta s}{d\tau^2} + 7\mathrm{m}^2 \,\delta s \cos 2\tau + 2\frac{d\delta p}{d\tau}\left(1 + \mathrm{m} - \tfrac{5}{4}\mathrm{m}^2 \cos 2\tau\right) - 2\mathrm{m}^2 \delta p \sin 2\tau = 0 \ldots (25).$$

This is the second of our equations transformed.

The Jacobian integral gives

$$\frac{d\delta s}{d\tau} + \delta p \frac{d\phi}{d\tau} = \frac{3\mathrm{m}^2 x \delta x}{V} - \frac{\kappa}{V r^3}\left[\delta p \left(x \cos \phi + y \sin \phi\right) + \delta s \left(-x \sin \phi + y \cos \phi\right)\right]$$

$$= 3\mathrm{m}^2 \cos \tau \left(\delta p \cos \tau - \delta s \sin \tau\right) - \left(1 + 2\mathrm{m} + \tfrac{3}{2}\mathrm{m}^2 - \tfrac{7}{4}\mathrm{m}^2 \cos 2\tau\right.$$

$$\left. + 3\mathrm{m}^2 \cos 2\tau\right)\left[\delta p \left(1 - \mathrm{m}^2 \cos 2\tau\right) + 2\mathrm{m}^2 \delta s \sin 2\tau\right]$$

$$= \frac{3m^2}{2} \delta p \left(1 + \cos 2\tau\right) - \frac{3m^2}{2} \delta s \sin 2\tau$$

$$- \delta p \left(1 + 2m + \tfrac{3}{2}m^2 + \tfrac{5}{4}m^2 \cos 2\tau - m^2 \cos 2\tau\right) - 2m^2 \delta s \sin 2\tau$$

$$= - \delta p \left(1 + 2m - \tfrac{5}{4}m^2 \cos 2\tau\right) - \tfrac{7}{2}m^2 \delta s \sin 2\tau.$$

Substituting for $\dfrac{d\phi}{d\tau}$ its value from (21)

$$\frac{d\,\delta s}{d\tau} = - \delta p \left(1 + 2m - \tfrac{5}{4}m^2 \cos 2\tau\right) - \delta p \left(1 - \tfrac{5}{4}m^2 \cos 2\tau\right) - \tfrac{7}{2}m^2 \delta s \sin 2\tau$$

$$= - \delta p \left(2 + 2m - \tfrac{5}{2}m^2 \cos 2\tau\right) - \tfrac{7}{2}m^2 \delta s \sin 2\tau$$

$$\frac{2d\,\delta s}{d\tau} = - 4\delta p \left(1 + m - \tfrac{5}{4}m^2 \cos 2\tau\right) - 7m^2 \delta s \sin 2\tau$$

$$\frac{2d\,\delta s}{d\tau} \left(1 + m - \tfrac{5}{4}m^2 \cos 2\tau\right) = - 4\delta p \left(1 + 2m + m^2 - \tfrac{5}{2}m^2 \cos 2\tau\right) - 7m^2 \delta s \sin 2\tau$$

$$\frac{2d\,\delta s}{d\tau} \left(1 + m - \tfrac{5}{4}m^2 \cos 2\tau\right) + 7m^2 \delta s \sin 2\tau = - 4\delta p \left(1 + 2m + m^2 - \tfrac{5}{2}m^2 \cos 2\tau\right)$$

$$\dots\dots(26).$$

This expression occurs in (24), and therefore can be used to eliminate $\dfrac{d\,\delta s}{d\tau}$ from it.

Substituting we get

$$\frac{d^2\delta p}{d\tau^2} + \delta p \left[- 3 - 6m - \tfrac{9}{2}m^2 - 5m^2 \cos 2\tau + 4 + 8m + 4m^2 - 10m^2 \cos 2\tau \right] = 0,$$

i.e.
$$\frac{d^2\delta p}{d\tau^2} + \delta p \left[1 + 2m - \tfrac{1}{2}m^2 - 15m^2 \cos 2\tau\right] = 0.$$

And
$$\frac{d\,\delta s}{d\tau} = - 2\delta p \left(1 + m - \tfrac{5}{4}m^2 \cos 2\tau\right) - \tfrac{7}{2}m^2 \delta s \sin 2\tau$$

$$\left. \right\} \dots\dots(27).$$

If we differentiate the second of these equations, which it is to be remembered was derived from Jacobi's integral and therefore involves our second differential equation, we get

$$\frac{d^2\delta s}{d\tau^2} + 7m^2 \delta s \cos 2\tau + \tfrac{7}{2}m^2 \sin 2\tau \frac{d\,\delta s}{d\tau} + 2\left(1 + m - \tfrac{5}{4}m^2 \cos 2\tau\right) \frac{d\,\delta p}{d\tau}$$

$$+ 5m^2 \delta p \sin 2\tau = 0,$$

and eliminating $\dfrac{d\,\delta s}{d\tau}$

$$\frac{d^2\delta s}{d\tau^2} + 7m^2 \delta s \cos 2\tau - 7m^2 \delta p \sin 2\tau + 2\left(1 + m - \tfrac{5}{4}m^2 \cos 2\tau\right) \frac{d\,\delta p}{d\tau}$$

$$+ 5m^2 \delta p \sin 2\tau = 0,$$

or $\qquad \dfrac{d^2 \delta s}{d\tau^2} + 7m^2 \delta s \cos 2\tau + 2\left(1 + m - \tfrac{5}{4}m^2 \cos 2\tau\right)\dfrac{d\delta p}{d\tau} - 2m^2 \delta p \sin 2\tau = 0,$

and this is as might be expected our second differential equation which was found above. Hence we only require to consider the equations (27).

§ 6. *Integration of an important type of Differential Equation.*

The differential equation for δp belongs to a type of great importance in mathematical physics. We may write the typical equation in the form

$$\dfrac{d^2 x}{dt^2} + \left(\Theta_0 + 2\Theta_1 \cos 2t + 2\Theta_2 \cos 4t + \ldots\right)x = 0,$$

where Θ_0, Θ_1, Θ_2, ... are constants depending on increasing powers of a small quantity m. It is required to find a solution such that x remains small for all values of t.

Let us attempt the apparently obvious process of solution by successive approximations.

Neglecting Θ_1, Θ_2, ..., we get as a first approximation

$$x = A \cos\left(t \sqrt{\Theta_0} + \epsilon\right).$$

Using this value for x in the term multiplied by Θ_1, and neglecting Θ_2, Θ_3, ..., we get

$$\dfrac{d^2 x}{dt^2} + \Theta_0 x + A\Theta_1 \left\{\cos\left[t\left(\sqrt{\Theta_0} + 2\right) + \epsilon\right] + \cos\left[t\left(\sqrt{\Theta_0} - 2\right) + \epsilon\right]\right\} = 0.$$

Solving this by the usual rules we get the second approximation

$$x = A\left\{\cos\left[t\sqrt{\Theta_0} + \epsilon\right] + \dfrac{\Theta_1 \cos\left[t\left(\sqrt{\Theta_0} + 2\right) + \epsilon\right]}{4\left(\sqrt{\Theta_0} + 1\right)} - \dfrac{\Theta_1 \cos\left[t\left(\sqrt{\Theta_0} - 2\right) + \epsilon\right]}{4\left(\sqrt{\Theta_0} - 1\right)}\right\}.$$

Again using this we have the differential equation

$$\dfrac{d^2 x}{dt^2} + \Theta_0 x + A\Theta_1 \left\{\cos\left[t\left(\sqrt{\Theta_0} + 2\right) + \epsilon\right] + \cos\left[t\left(\sqrt{\Theta_0} - 2\right) - \epsilon\right]\right\}$$

$$+ \dfrac{A\Theta_1^2}{4\left(\sqrt{\Theta_0} + 1\right)}\left\{\cos\left[t\left(\sqrt{\Theta_0} + 4\right) + \epsilon\right] + \cos\left(t\sqrt{\Theta_0} + \epsilon\right)\right\}$$

$$- \dfrac{A\Theta_1^2}{4\left(\sqrt{\Theta_0} - 1\right)}\left\{\cos\left(t\sqrt{\Theta_0} + \epsilon\right) + \cos\left[t\left(\sqrt{\Theta_0} - 4\right) + \epsilon\right]\right\}$$

$$+ A\Theta_2 \left\{\cos\left[t\left(\sqrt{\Theta_0} + 4\right) + \epsilon\right] + \cos\left[t\left(\sqrt{\Theta_0} - 4\right) + \epsilon\right]\right\} = 0.$$

Now this equation involves terms of the form $B \cos\left(t\sqrt{\Theta_0} + \epsilon\right)$; on integration terms of the form $Ct \sin\left(t\sqrt{\Theta_0} + \epsilon\right)$ will arise. But these terms are not periodic and do not remain small when t increases. x will therefore not remain small and the argument will fail. The assumption on which these approximations have been made is that the period of the principal term of x can be determined from Θ_0 alone and is independent of Θ_1, Θ_2, ... But the appearance of secular terms leads us to revise this assumption and to take as a first approximation

$$x = A \cos (ct \sqrt{\Theta_0} + \epsilon),$$

where c is nearly equal to 1 and will be determined, if possible, to prevent secular terms arising.

It will, however, be more convenient to write as a first approximation

$$x = A \cos (ct + \epsilon),$$

where c is nearly equal to $\sqrt{\Theta_0}$.

Using this value of x in the term involving Θ_1, our equation becomes

$$\frac{d^2x}{dt^2} + \Theta_0 x + A\Theta_1 \{\cos [(c+2) t + \epsilon] + \cos [(c-2) t + \epsilon]\} = 0,$$

and the second approximation is

$$x = A \cos (ct + \epsilon) + \frac{A\Theta_1}{(c+2)^2 - \Theta_0} \cos [(c+2) t + \epsilon]$$
$$+ \frac{A\Theta_1}{(c-2)^2 - \Theta_0} \cos [(c-2) t + \epsilon].^*$$

Proceeding to another approximation with this value of x, we get

$$\frac{d^2x}{dt^2} + \Theta_0 x + A\Theta_1 \{\cos [(c+2) t + \epsilon] + \cos [(c-2) t + \epsilon]\}$$
$$+ \frac{A\Theta_1{}^2}{(c+2)^2 - \Theta_0} \{\cos [(c+4) t + \epsilon] + \cos (ct + \epsilon)\}$$
$$+ \frac{A\Theta_1{}^2}{(c-2)^2 - \Theta_0} \{\cos (ct + \epsilon) + \cos [(c-4) t + \epsilon]\}$$
$$+ A\Theta_2 \{\cos [(c+4) t + \epsilon] + \cos [(c-4) t + \epsilon]\} = 0.$$

We might now proceed to further approximations but just as a term in $\cos (ct + \epsilon)$ generates in the solution terms in

$$\cos [(c \pm 2) t + \epsilon] \quad \text{and} \quad \cos [(c \pm 4) t + \epsilon],$$

terms in $\quad \cos [(c \pm 2) t + \epsilon] \quad \text{and} \quad \cos [(c \pm 4) t + \epsilon]$

will generate new terms in $\cos (ct + \epsilon)$, i.e. terms of exactly the same nature as the term initially assumed. Hence to get our result it will be best to begin by assuming a series containing all the terms which will arise.

Various writers have found it convenient to introduce exponential instead of trigonometric functions. Following their example we shall therefore write the differential equation in the form

$$\frac{d^2x}{dt^2} + x \sum_{-\infty}^{+\infty} \Theta_i e^{2it \sqrt{-1}} = 0 \quad \dots\dots\dots\dots(28),$$

where $\qquad \Theta_{-i} = \Theta_i,$

* This is not a solution of the previous equation, unless we actually put $c = \sqrt{\Theta_0}$ in the first term.

and the solution is assumed to be

$$x = \sum_{-\infty}^{+\infty} A_j e^{(c+2j)\,t\sqrt{-1}},$$

where the ratios of all the coefficients A_j, and c, are to be determined by equating coefficients of different powers of $e^{t\sqrt{-1}}$.

Substituting this expression for x in the differential equation, we get

$$-\sum_{-\infty}^{+\infty} (c+2j)^2 A_j e^{(c+2j)\,t\sqrt{-1}} + \sum_{-\infty}^{+\infty} A_j e^{(c+2j)\,t\sqrt{-1}} \sum_{-\infty}^{+\infty} \Theta_i e^{2it\sqrt{-1}} = 0,$$

and equating to zero the coefficient of $e^{(c+2j)\,t\sqrt{-1}}$,

$$-(c+2j)^2 A_j + A_j\Theta_0 + A_{j-1}\Theta_1 + A_{j-2}\Theta_2 + A_{j-3}\Theta_3 + \ldots$$
$$+ A_{j+1}\Theta_{-1} + A_{j+2}\Theta_{-2} + A_{j+3}\Theta_{-3} + \ldots = 0.$$

Hence the succession of equations is

$$\ldots \ldots \ldots \ldots \ldots \ldots \ldots \ldots \ldots \ldots \ldots \ldots \ldots \ldots \ldots \ldots$$

$$\ldots + [\Theta_0 - (c-4)^2] A_{-2} + \Theta_{-1}A_{-1} + \Theta_{-2}A_0 + \Theta_{-3}A_1 + \Theta_{-4}A_2 + \ldots = 0,$$

$$\ldots + \Theta_1 A_{-2} + [\Theta_0 - (c-2)^2] A_{-1} + \Theta_{-1}A_0 + \Theta_{-2}A_1 + \Theta_{-3}A_2 + \ldots = 0,$$

$$\ldots + \Theta_2 A_{-2} + \Theta_1 A_{-1} + (\Theta_0 - c^2) A_0 + \Theta_{-1}A_1 + \Theta_{-2}A_2 + \ldots = 0,$$

$$\ldots + \Theta_3 A_{-2} + \Theta_2 A_{-1} + \Theta_1 A_0 + [\Theta_0 - (c+2)^2] A_1 + \Theta_{-1}A_2 + \ldots = 0,$$

$$\ldots + \Theta_4 A_{-2} + \Theta_3 A_{-1} + \Theta_2 A_0 + \Theta_1 A_1 + [\Theta_0 - (c+4)^2] A_2 + \ldots = 0.$$

$$\ldots \ldots \ldots \ldots \ldots \ldots \ldots \ldots \ldots \ldots \ldots \ldots \ldots \ldots \ldots \ldots$$

We clearly have an infinite determinantal equation for c.

If we take only three columns and rows, we get

$$[\Theta_0 - (c-2)^2][\Theta_0 - c^2][\Theta_0 - (c+2)^2] - \Theta_1^2[\Theta_0 - (c-2)^2] - \Theta_1^2[\Theta_0 - (c+2)^2]$$
$$- \Theta_2^2(\Theta_0 - c^2) + 2\Theta_1^2\Theta_2 = 0,$$

$$[(\Theta_0 - c^2 - 4)^2 - 16c^2][\Theta_0 - c^2] - 2\Theta_1^2(\Theta_0 - c^2 - 4) - \Theta_2^2(\Theta_0 - c^2) + 2\Theta_1^2\Theta_2 = 0.$$

If we neglect $(\Theta_0 - c^2)^3$ which is certainly small

$$[-8(\Theta_0 - c^2) + 16 + 16(\Theta_0 - c^2) - 16\Theta_0][\Theta_0 - c^2]$$
$$- (\Theta_0 - c^2)[2\Theta_1^2 + \Theta_2^2] + 8\Theta_1^2 + 2\Theta_1^2\Theta_2 = 0,$$

$$8(\Theta_0 - c^2)^2 + (\Theta_0 - c^2)(16 - 16\Theta_0 - 2\Theta_1^2 - \Theta_2^2) + 8\Theta_1^2 + 2\Theta_1^2\Theta_2 = 0,$$

$$(\Theta_0 - c^2)^2 + 2(\Theta_0 - c^2)(1 - \Theta_0 - \tfrac{1}{8}\Theta_1^2 - \tfrac{1}{16}\Theta_2^2) + \Theta_1^2 + \tfrac{1}{4}\Theta_1^2\Theta_2 = 0.$$

Since Θ_1^2, Θ_2^2 are small compared with $1 - \Theta_0$, and Θ_2 compared with 1, we have as a rougher approximation

$$(c^2 - \Theta_0)^2 + 2(\Theta_0 - 1)(c^2 - \Theta_0) = -\Theta_1^2,$$

whence
$$c^2 - \Theta_0 = -(\Theta_0 - 1) \pm \sqrt{(\Theta_0 - 1)^2 - \Theta_1{}^2},$$
$$c^2 = 1 \pm \sqrt{(\Theta_0 - 1)^2 - \Theta_1{}^2}.$$

Now $c^2 = \Theta_0$ when $\Theta_1 = 0$. Hence we take the positive sign and get
$$c = \sqrt{1 + \sqrt{(\Theta_0 - 1)^2 - \Theta_1{}^2}} \quad \dots\dots\dots\dots\dots\dots(29),$$

which is wonderfully nearly correct.

For further discussion of the equation for c, see Notes 1, 2, pp. 53, 55.

§ 7. *Integration of the Equation for δp.*

We now return to the Lunar Theory and consider the solution of our differential equation. Assume it to be
$$\delta p = A_{-1} \cos\left[(c-2)\,\tau + \epsilon\right] + A_0 \cos\left(c\tau + \epsilon\right) + A_1 \cos\left[(c+2)\,\tau + \epsilon\right].$$

On substitution in (27) we get
$$A_{-1}\left[(1 + 2m - \tfrac{1}{2}m^2 - 15m^2 \cos 2\tau) - (c-2)^2\right] \cos\left[(c-2)\,\tau + \epsilon\right]$$
$$+ A_0 \ \left[(1 + 2m - \tfrac{1}{2}m^2 - 15m^2 \cos 2\tau) - c^2\right] \cos\left(c\tau + \epsilon\right)$$
$$+ A_1 \ \left[(1 + 2m - \tfrac{1}{2}m^2 - 15m^2 \cos 2\tau) - (c+2)^2\right] \cos\left[(c+2)\,\tau + \epsilon\right] = 0.$$

Then we equate to zero the coefficients of the several cosines.

1st $\cos(c\tau + \epsilon)$ gives
$$-\tfrac{15}{2}m^2 A_{-1} + A_0(1 + 2m - \tfrac{1}{2}m^2 - c^2) - \tfrac{15}{2}m^2 A_1 = 0.$$

2nd $\cos\left[(c-2)\,\tau + \epsilon\right]$ gives
$$A_{-1}\left[1 + 2m - \tfrac{1}{2}m^2 - (c-2)^2\right] - \tfrac{15}{2}m^2 A_0 = 0.$$

3rd $\cos\left[(c+2)\,t + \epsilon\right]$ gives
$$-\tfrac{15}{2}m^2 A_0 + A_1\left[1 + 2m - \tfrac{1}{2}m^2 - (c+2)^2\right] = 0.$$

If we neglect terms in m^2 the first equation gives us $c^2 = 1 + 2m$, and therefore $c = 1 + m,\ c - 2 = -(1 - m),\ c + 2 = 3 + m$.

The second and third equations then reduce to
$$4m A_{-1} = 0; \quad A_1(-8 - 4m) = 0.$$

From this it follows that A_{-1} is at least of order m and A_1 at least of order m^2.

Then since we are neglecting higher powers than m^2, the first equation reduces to
$$A_0(1 + 2m - \tfrac{1}{2}m^2 - c^2) = 0,$$

so that
$$c^2 = 1 + 2m - \tfrac{1}{2}m^2 \quad \text{or} \quad c = 1 + m - \tfrac{3}{4}m^2.$$

Thus
$$(c - 2)^2 = (1 - m + \tfrac{3}{4}m^2)^2 = 1 - 2m + \tfrac{5}{2}m^2,$$

and
$$1 + 2m - \tfrac{1}{2}m^2 - (c - 2)^2 = 4m - 3m^2.$$

Hence the second equation becomes

$$A_{-1}(4\mathrm{m} - 3\mathrm{m}^2) = \tfrac{15}{2}\mathrm{m}^2 A_0;$$

and since A_{-1} is of order m, the term $-3\mathrm{m}^2 A_{-1}$ is of order m^3 and therefore negligible.   Hence

$$4\mathrm{m}\, A_{-1} = \tfrac{15}{2}\mathrm{m}^2 A_0 \quad \text{or} \quad A_{-1} = \tfrac{15}{8}\mathrm{m}A_0,$$

and we cannot obtain A_{-1} to an order higher than the first.

The third equation is

$$- \tfrac{15}{2}\mathrm{m}^2 A_0 + A_1\,[1 - 9] = 0,$$

or $$A_1 = - \tfrac{15}{16}\mathrm{m}^2 A_0.$$

We have seen that A_{-1} can only be obtained to the first order; so it is useless to retain terms of a higher order in A_1.   Hence our solution is

$$A_{-1} = \tfrac{15}{8}\mathrm{m}A_0, \quad A_1 = 0.$$

Hence $$\delta p = A_0 \left\{\cos (c\tau + \epsilon) + \tfrac{15}{8}\mathrm{m} \cos \left[(c - 2)\,\tau + \epsilon\right]\right\} \dots\dots\dots(30).$$

In order that the solution may agree with the more ordinary notation we write $A_0 = -\, a_0 e$, and obtain

$$\left. \begin{aligned} \delta p &= -\, a_0 e \cos (c\tau + \epsilon) - \tfrac{15}{8}\mathrm{m}a_0 e \cos \left[(c - 2)\,\tau + \epsilon\right] \\ c &= 1 + \mathrm{m} - \tfrac{3}{4}\mathrm{m}^2 \end{aligned} \right\} \ \ \dots\dots(31).$$

where

To the first order of small quantities the equation (27) for δs was

$$\frac{d\,\delta s}{d\tau} = -\,2\,(1 + \mathrm{m})\,\delta p$$

$$= 2\,(1 + \mathrm{m})\, a_0 e \cos (c\tau + \epsilon) + \tfrac{15}{4}\mathrm{m}a_0 e \cos \left[(c - 2)\,\tau + \epsilon\right].$$

If we integrate and note that $c = 1 + \mathrm{m}$ so that $c - 2 = -(1 - \mathrm{m})$, we have

$$\delta s = 2 a_0 e \sin (c\tau + \epsilon) - \tfrac{15}{4}\mathrm{m}a_0 e \sin \left[(c - 2)\,\tau + \epsilon\right] \ \ \dots\dots(32).$$

We take the constant of integration zero because $e = 0$ will then correspond to no displacement along the variational curve.

In order to understand the physical meaning of the results let us consider the solution when $\mathrm{m} = 0$, i.e. when the solar perturbation vanishes.

Then $$\delta p = -\, a_0 e \cos (c\tau + \epsilon), \quad \delta s = 2 a_0 e \sin (c\tau + \epsilon).$$

In the undisturbed orbit

$$x = a_0 \cos \tau, \quad y = a_0 \sin \tau \ \text{ so that } \ \phi = \tau,$$

and $$\delta x = \delta p \cos \phi - \delta s \sin \phi,$$

$$\delta y = \delta p \sin \phi + \delta s \cos \phi\,;$$

$$\delta x = -\, a_0 e \cos (c\tau + \epsilon) \cos \tau - 2 a_0 e \sin (c\tau + \epsilon) \sin \tau,$$

$$\delta y = -\, a_0 e \cos (c\tau + \epsilon) \sin \tau + 2 a_0 e \sin (c\tau + \epsilon) \cos \tau.$$

Therefore writing $X = x + \delta x$, $Y = y + \delta y$, $X = R \cos \Theta$, $Y = R \sin \Theta$,

$$X = a_0 \left[\cos \tau - e \cos (c\tau + \epsilon) \cos \tau - 2e \sin (c\tau + \epsilon) \sin \tau \right],$$

$$Y = a_0 \left[\sin \tau - e \cos (c\tau + \epsilon) \sin \tau + 2e \sin (c\tau + \epsilon) \cos \tau \right].$$

Therefore $\qquad\qquad R^2 = a_0{}^2 \left[1 - 2e \cos (c\tau + \epsilon) \right]$

or $\qquad\qquad R = a_0 \left[1 - e \cos (c\tau + \epsilon) \right] = \dfrac{a_0}{1 + e \cos (c\tau + \epsilon)}$ ....(33).

Again $\qquad\qquad \cos \Theta = \cos \tau - 2e \sin (c\tau + \epsilon) \sin \tau,$

$\qquad\qquad\qquad \sin \Theta = \sin \tau + 2e \sin (c\tau + \epsilon) \cos \tau.$

Hence $\qquad\qquad \sin (\Theta - \tau) = 2e \sin (c\tau + \epsilon),$

giving $\qquad\qquad \Theta = \tau + 2e \sin (c\tau + \epsilon)$ ....(34).

It will be noted that the equations for R, Θ are of the same form as the first approximation to the radius vector and true longitude in undisturbed elliptic motion. When we neglect the solar perturbation by putting $\mathrm{m} \doteq 0$ we see that e is to be identified with the eccentricity and $c\tau + \epsilon$ with the mean anomaly.

* We can interpret c in terms of the symbols of the ordinary lunar theories. When no perturbations are considered the moon moves in an ellipse. The perturbations cause the moon to deviate from this simple path. If a fixed ellipse is taken, these deviations increase with the time. It is found, however, that if we consider the ellipse to be fixed in shape and size but with the line of apses moving with uniform angular velocity, the actual motion of the moon differs from this modified elliptic motion only by small periodic quantities. If n denote as before the mean sidereal motion of the moon and $\dfrac{d\varpi}{dt}$ the mean motion of the line of apses, the argument entering into the elliptic inequalities is $\left(n - \dfrac{d\varpi}{dt} \right) t + \epsilon$. This must be the same as $c\tau + \epsilon$, i.e. as $c (n - n') t + \epsilon$.

Hence $\qquad\qquad n - \dfrac{d\varpi}{dt} = c (n - n'),$

giving $\qquad\qquad \dfrac{d\varpi}{n\,dt} = 1 - c \dfrac{n - n'}{n}$

$$= 1 - \frac{c}{1 + \mathrm{m}} \quad \text{since} \quad \mathrm{m} = \frac{n'}{n - n'}.$$

A determination of c is therefore equivalent to a determination of the rate of change of perigee; the value of c we have already obtained gives

$$\frac{d\varpi}{n\,dt} = \tfrac{3}{4}\mathrm{m}^2.$$

---

* From here till the foot of this page a slight knowledge of ordinary lunar theory is supposed. The results given are not required for the further development of Hill's theory.

Returning to our solution, and for simplicity again dropping the factor a_0, we have from (31), (32)

$$\delta p = -\tfrac{15}{8}me\cos\left[(c-2)\,\tau+\epsilon\right] - e\cos(c\tau+\epsilon),$$

$$\delta s = -\tfrac{15}{4}me\sin\left[(c-2)\,\tau+\epsilon\right] + 2e\sin(c\tau+\epsilon).$$

Also $\cos\phi = \cos\tau$, $\sin\phi = \sin\tau$ to the first order of small quantities, and

$$\delta x = \delta p\cos\phi - \delta s\sin\phi,\quad \delta y = \delta p\sin\phi + \delta s\cos\phi.$$

Therefore

$$\delta x = -\tfrac{15}{8}me\cos\left[(c-2)\,\tau+\epsilon\right]\cos\tau - e\cos(c\tau+\epsilon)\cos\tau$$
$$+ \tfrac{15}{4}me\sin\left[(c-2)\,\tau+\epsilon\right]\sin\tau - 2e\sin(c\tau+\epsilon)\sin\tau,$$

$$\delta y = -\tfrac{15}{8}me\cos\left[(c-2)\,\tau+\epsilon\right]\sin\tau - e\cos(c\tau+\epsilon)\sin\tau$$
$$- \tfrac{15}{4}me\sin\left[(c-2)\,\tau+\epsilon\right]\cos\tau + 2e\sin(c\tau+\epsilon)\cos\tau.$$

Now let $X = x + \delta x$, $Y = y + \delta y$ and we have by means of the values of x, y in the variational curve

$$X = \cos\tau\left[1 - m^2 - \tfrac{3}{4}m^2\sin^2\tau - \tfrac{15}{8}me\cos\{(c-2)\,\tau+\epsilon\} - e\cos(c\tau+\epsilon)\right]$$
$$+ \sin\tau\left[\tfrac{15}{4}me\sin\{(c-2)\,\tau+\epsilon\} - 2e\sin(c\tau+\epsilon)\right],$$

$$Y = \sin\tau\left[1 + m^2 + \tfrac{3}{4}m^2\cos^2\tau - \tfrac{15}{8}me\cos\{(c-2)\,\tau+\epsilon\} - e\cos(c\tau+\epsilon)\right]$$
$$- \cos\tau\left[\tfrac{15}{4}me\sin\{(c-2)\,\tau+\epsilon\} - 2e\sin(c\tau+\epsilon)\right].$$

Writing $R^2 = X^2 + Y^2$, we obtain to the requisite degree of approximation

$$R^2 = \cos^2\tau\left[1 - 2m^2 - \tfrac{3}{2}m^2\sin^2\tau - \tfrac{15}{4}me\cos\{(c-2)\,\tau+\epsilon\} - 2e\cos(c\tau+\epsilon)\right]$$

$$+ \sin^2\tau\left[1 + 2m^2 + \tfrac{3}{2}m^2\cos^2\tau - \tfrac{15}{5}me\cos\{(c-2)\,\tau+\epsilon\} - 2e\cos(c\tau+\epsilon)\right]$$

$$+ \sin 2\tau\left[\tfrac{15}{4}me\sin\{(c-2)\,\tau+\epsilon\} - 2e\sin(c\tau+\epsilon)\right]$$

$$- \sin 2\tau\left[\tfrac{15}{4}me\sin\{(c-2)\,\tau+\epsilon\} - 2e\sin(c\tau+\epsilon)\right],$$

$$R^2 = 1 - 2m^2\cos 2\tau - \tfrac{15}{4}me\cos\{(c-2)\,\tau+\epsilon\} - 2e\cos(c\tau+\epsilon).$$

Hence reintroducing the factor a_0 which was omitted for the sake of brevity

$$R = a_0\left[1 - e\cos(c\tau+\epsilon) - \tfrac{15}{8}me\cos\{(c-2)\,\tau+\epsilon\} - m^2\cos 2\tau\right]\ldots(35).$$

This gives the radius vector; it remains to find the longitude.

We multiply the expressions for X, Y by $1/R$, i.e. by

$$1 + e\cos(c\tau+\epsilon) + \tfrac{15}{8}me\cos\left[(c-2)\,\tau+\epsilon\right] + m^2\cos 2\tau,$$

and remembering that

$$m^2\cos 2\tau = m^2 - 2m^2\sin^2\tau = 2m^2\cos^2\tau - m^2,$$

we get

$$\cos\Theta = \cos\tau\left[1 - \tfrac{11}{4}m^2\sin^2\tau\right] + \sin\tau\left[\tfrac{15}{4}me\sin\{(c-2)\,\tau+\epsilon\} - 2e\sin(c\tau+\epsilon)\right],$$

$$\sin\Theta = \sin\tau\left[1 + \tfrac{11}{4}m^2\cos^2\tau\right] - \cos\tau\left[\tfrac{15}{4}me\sin\{(c-2)\,\tau+\epsilon\} - 2e\sin(c\tau+\epsilon)\right].$$

Whence

$$\sin(\Theta - \tau) = \tfrac{11}{8}m^2\sin 2\tau - \tfrac{15}{4}me\sin\{(c-2)\,\tau+\epsilon\} + 2e\sin(c\tau+\epsilon),$$

or to our degree of approximation

$$\Theta = \tau + \tfrac{11}{8} m^2 \sin 2\tau - \tfrac{15}{4} me \sin \{(c - 2)\tau + \epsilon\} + 2e \sin (c\tau + \epsilon)...(36).$$

We now transform these results into the ordinary notation.

* Let l, v be the moon's mean and true longitudes, and l' the sun's mean longitude. Then Θ being the moon's true longitude relatively to the moving axes, we have

$$v = \Theta + l'.$$

Also
$$\tau + l' = (n - n')t + n't = l,$$
$$\therefore \ \tau = l - l'.$$

We have seen that $c\tau + \epsilon$ is the moon's mean anomaly, or $l - \varpi$,

$$\therefore \ (c - 2)\tau + \epsilon = l - \varpi - 2(l - l') = -(l + \varpi - 2l').$$

Then substituting these values in the expressions for R and Θ and adding l' to the latter we have on noting that $a_0 = a(1 - \tfrac{1}{6}m^2)$

$$R = a\left[1 - \tfrac{1}{6}m^2 - e\cos(l - \varpi) - \tfrac{15}{8}me\cos(l - 2l' + \varpi) - m^2\cos 2(l - l')\right]$$
$$\text{equation of centre} \qquad \text{evection} \qquad\qquad \cdot\text{variation}$$
$$v = l + 2e\sin(l - \varpi) + \tfrac{15}{4}me\sin(l - 2l' + \varpi) + \tfrac{11}{8}m^2\sin 2(l - l')$$
$$\text{equation of centre} \qquad \text{evection} \qquad\qquad \text{variation}$$
$$.........(37).$$

The names of the inequalities in radius vector and longitude are written below, and the values of course agree with those found in ordinary lunar theories.

§ 8. *Introduction of the Third Coordinate.*

Still keeping $\Omega = 0$, consider the differential equation for z in (5)

$$\frac{d^2 z}{d\tau^2} + \frac{\kappa z}{r^3} + m^2 z = 0.$$

From (8)
$$\frac{\kappa}{a_0^3} = 1 + 2m + \tfrac{3}{2}m^2,$$

and from (10)
$$\frac{a_0^3}{r^3} = 1 + 3m^2\cos 2\tau.$$

The equation may therefore be written

$$\frac{d^2 z}{d\tau^2} + z(1 + 2m + \tfrac{5}{2}m^2 + 3m^2\cos 2\tau) = 0.$$

This is an equation of the type considered in § 6 and therefore we assume

$$z = B_{-1}\cos\{(g - 2)\tau + \zeta\} + B_0\cos(g\tau + \zeta) + B_1\cos\{(g + 2)\tau + \zeta\}.$$

* From here till the end of this paragraph is not a part of Hill's theory, it is merely a comparison with ordinary lunar theories.

On substitution we get

$$B_{-1}[-(g-2)^2+1+2m+\tfrac{5}{2}m^2+3m^2\cos 2\tau]\cos[(g-2)\tau+\zeta]$$

$$+ B_0 \ [-g^2+1+2m+\tfrac{5}{2}m^2+3m^2\cos 2\tau]\cos(g\tau+\zeta)$$

$$+ B_1 \ [-(g+2)^2+1+2m+\tfrac{5}{2}m^2+3m^2\cos 2\tau]\cos[(g+2)\tau+\zeta]=0.$$

The coefficients of $\cos(g\tau+\zeta)$, $\cos[(g-2)\tau+\zeta]$, $\cos[(g+2)\tau+\zeta]$ give respectively

$$\left.\begin{aligned}
\tfrac{3}{2}m^2 B_{-1}+B_0[-g^2+1+2m+\tfrac{5}{2}m^2]+\tfrac{3}{2}m^2 B_1 &= 0\\
B_{-1}[-(g-2)^2+1+2m+\tfrac{5}{2}m^2]+\tfrac{3}{2}m^2 B_0 \quad &= 0\\
\tfrac{3}{2}m^2 B_0+B_1[-(g+2)^2+1+2m+\tfrac{5}{2}m^2] \quad &= 0
\end{aligned}\right\}\dots\dots(38).$$

As a first approximation drop the terms in m^2. The first of these equations then gives $g^2=1+2m$. The third equation then shews that $\dfrac{B_1}{B_0}$ is of order m^2. But a factor m can be removed from the second equation shewing that $\dfrac{B_{-1}}{B_0}$ is of order m and can only be determined to this order. Hence B_1 can be dropped. [Cf. pp. 39, 40.]

Considering terms in m^2 we now get from the first equation

$$g^2=1+2m+\tfrac{5}{2}m^2.$$

Therefore $\qquad g=1+m+\tfrac{5}{4}m^2-\tfrac{1}{2}m^2=1+m+\tfrac{3}{4}m^2,$

$$(g-2)^2=(1-m)^2=1-2m,\text{ neglecting terms in }m^2.$$

The second equation then gives

$$B_{-1}=-\tfrac{3}{8}m\,B_0,$$

and the solution is

$$z=B_0[\cos(g\tau+\zeta)-\tfrac{3}{8}m\cos\{(g-2)\tau+\zeta\}]\ \dots\dots\dots(39).$$

We shall now interpret this equation geometrically. To do so we neglect the solar perturbation and we get

$$z=B_0\cos(g\tau+\zeta)\ \dots\dots\dots\dots\dots(40).$$

Now consider the moon to move in a plane orbit inclined at angle i to the ecliptic and let Ω be the longitude of the lunar node, l the longitude of the moon, β the latitude.

The right-angled spherical triangle gives

$$\tan\beta=\tan i\sin(l-\Omega)$$

and therefore

$$z=r\tan\beta=r\tan i\sin(l-\Omega).$$

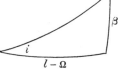

Fig. 3.

As we are only dealing with a first approximation we may put $r = a_0$ and so we interpret

$$B_0 = a_0 \tan i,$$
$$g\tau + \zeta = l - \Omega - \tfrac{1}{2}\pi.$$

* We can easily find the significance of g, for differentiating this equation with respect to the time we get

$$g(n - n') = n - \frac{d\Omega}{dt},$$

$$\therefore \frac{d\Omega}{ndt} = 1 - \frac{g(n - n')}{n}$$

$$= 1 - \frac{g}{1 + m}$$

$$= -\tfrac{3}{4}m^2 \text{ to our approximation.}$$

Thus we find that the node has a retrograde motion.

We have
$$g\tau + \zeta = l - \Omega - \tfrac{1}{2}\pi,$$
$$(g - 2)\tau + \zeta = l - \Omega - \tfrac{1}{2}\pi - 2(l - l')$$
$$= -(l - 2l' + \Omega) - \tfrac{1}{2}\pi.$$

If we write $s = \tan\beta$, $k = \tan i$, we find

$$s = k\sin(l - \Omega) + \tfrac{3}{8}mk\sin(l - 2l' + \Omega) \quad \dots\dots\dots\dots(41).$$

The last term in this equation is called the evection in latitude.

§ 9. *Results obtained.*

We shall now shortly consider the progress we have made towards the actual solution of the moon's motion. We obtained first of all a special solution of the differential equations assuming the motion to be in the ecliptic and neglecting certain terms in the force function denoted by Ω†. This gave us a disturbed circular orbit in the plane of the ecliptic. We have since introduced the first approximation to two free oscillations about this motion, the one corresponding to eccentricity of the orbit, the other to an inclination of the orbit to the ecliptic.

It is found to be convenient to refer the motion of the moon to the projection on the ecliptic. We will denote by r_1 the curtate radius vector, so that $r_1^2 = x^2 + y^2$, $r^2 = r_1^2 + z^2$; the x, y axes rotating as before with angular velocity n' in the plane of the ecliptic. In determining the variational curve, § 3, we put $\Omega = 0$, $r = r_1$. It will appear therefore that in finding the actual motion of the moon we shall require to consider not only Ω but new terms in z^2. In the next section we shall discuss the actual motion of the moon, making use of the approximations we have already obtained.

* From here till end of paragraph is a comparison with ordinary lunar theories.
† The Ω of p. 20, not that of the preceding paragraph.

§ 10. *General Equations of Motion and their solution.*

We have $\qquad r_1{}^2 = x^2 + y^2$ and $r^2 = r_1{}^2 + z^2$.

Hence $\qquad \dfrac{1}{r^3} = \dfrac{1}{r_1{}^3}\left(1 - \dfrac{3}{2}\dfrac{z^2}{r_1{}^2}\right)$; and $\dfrac{1}{r} = \dfrac{1}{r_1}\left(1 - \dfrac{1}{2}\dfrac{z^2}{r_1{}^2}\right)$,

to our order of accuracy.

The original equations (3) may now be written

$$\left.\begin{aligned}
\frac{d^2x}{d\tau^2} - 2\mathrm{m}\frac{dy}{d\tau} + \frac{\kappa x}{r_1{}^3} - 3\mathrm{m}^2 x &= \frac{\partial\Omega}{\partial x} + \frac{3}{2}\frac{\kappa z^2 x}{r_1{}^5} \\[2mm]
\frac{d^2y}{d\tau^2} + 2\mathrm{m}\frac{dx}{d\tau} + \frac{\kappa y}{r_1{}^3} \qquad\quad &= \frac{\partial\Omega}{\partial y} + \frac{3}{2}\frac{\kappa z^2 y}{r_1{}^5} \\[2mm]
\frac{d^2z}{d\tau^2} \qquad\quad + \frac{\kappa z}{r_1{}^3} + \mathrm{m}^2 z \quad &= \frac{\partial\Omega}{\partial z} + \frac{3}{2}\frac{\kappa z^3}{r_1{}^5}
\end{aligned}\right\}\ \dots\dots\dots(42).$$

If we multiply by $2\dfrac{dx}{d\tau}$, $2\dfrac{dy}{d\tau}$, $2\dfrac{dz}{d\tau}$ and add, we find that the Jacobian integral becomes

$$V^2 = 2\frac{\kappa}{r_1} + \mathrm{m}^2\left(3x^2 - z^2\right) - \frac{\kappa z^2}{r_1{}^3} + 2\int_0^\tau\left(\frac{\partial\Omega}{\partial x}\frac{dx}{d\tau} + \frac{\partial\Omega}{\partial y}\frac{dy}{d\tau} + \frac{\partial\Omega}{\partial z}\frac{dz}{d\tau}\right)d\tau + C\dots(43),$$

where $\qquad V^2 = V_1{}^2 + \left(\dfrac{dz}{d\tau}\right)^2 = \left(\dfrac{dx}{d\tau}\right)^2 + \left(\dfrac{dy}{d\tau}\right)^2 + \left(\dfrac{dz}{d\tau}\right)^2.$

Now $\qquad \Omega = \tfrac{3}{2}\mathrm{m}^2\left(\dfrac{a'^3}{r'^3}r^2\cos^2\theta - x^2\right) + \tfrac{1}{2}\mathrm{m}^2 r^2\left(1 - \dfrac{a'^3}{r'^3}\right),$

and $\qquad \cos\theta = \dfrac{xx' + yy' + zz'}{rr'} = \dfrac{xx' + yy'}{rr'}$, since $z' = 0$.

Hence

$$\Omega = \tfrac{3}{2}\mathrm{m}^2\left\{\dfrac{a'^3}{r'^5}(xx' + yy')^2 - x^2\right\} + \tfrac{1}{2}\mathrm{m}^2(x^2 + y^2)\left(1 - \dfrac{a'^3}{r'^3}\right) + \tfrac{1}{2}\mathrm{m}^2 z^2\left(1 - \dfrac{a'^3}{r'^3}\right).$$

When we neglected Ω and z, we found the solution

$$x = a_0\left[(1 - \tfrac{19}{16}\mathrm{m}^2)\cos\tau + \tfrac{3}{16}\mathrm{m}^2\cos 3\tau\right],$$

$$y = a_0\left[(1 + \tfrac{19}{16}\mathrm{m}^2)\sin\tau + \tfrac{3}{16}\mathrm{m}^2\sin 3\tau\right].$$

We now require to determine the effect of the terms introduced on the right, and for brevity we write

$$X = \frac{\partial\Omega}{\partial x} + \frac{3}{2}\frac{\kappa z^2 x}{r_1{}^5}, \quad Y = \frac{\partial\Omega}{\partial y} + \frac{3}{2}\frac{\kappa z^2 y}{r_1{}^5}.$$

When we refer to § 4 and consider how the differential equations for δp, δs were formed from those for δx, δy, we see that the new terms X, Y on the right-hand sides of the differential equations for δx, δy will lead to new terms $X\cos\phi + Y\sin\phi$, $-X\sin\phi + Y\cos\phi$ on the right-hand sides of those for δp, δs.

Hence taking the equations (24) and (25) for δp and δs and introducing these new terms, we find

$$\frac{d^2\delta p}{d\tau^2} + \delta p\left[-3 - 6\mathrm{m} - \tfrac{9}{2}\mathrm{m}^2 - 5\mathrm{m}^2\cos 2\tau\right] - 2\frac{d\delta s}{d\tau}(1 + \mathrm{m} - \tfrac{5}{4}\mathrm{m}^2\cos 2\tau)$$

$$- 7\mathrm{m}^2\,\delta s\,\sin 2\tau = X\cos\phi + Y\sin\phi,$$

$$\frac{d^2\delta s}{d\tau^2} + 7\mathrm{m}^2\,\delta s\cos 2\tau + 2\frac{d\delta p}{d\tau}(1 + \mathrm{m} - \tfrac{5}{4}\mathrm{m}^2\cos 2\tau) - 2\mathrm{m}^2\,\delta p\,\sin 2\tau$$

$$= -X\sin\phi + Y\cos\phi.$$

In this analysis we shall include all terms to the order $\mathrm{m}k^2$, where k is the small quantity in the expression for z. Terms involving $\mathrm{m}^2 z^2$ will therefore be neglected. In the variation of the Jacobian integral the term $\dfrac{dz}{d\tau}\dfrac{d\delta z}{d\tau}$ can obviously be neglected. The variation of the Jacobian integral therefore gives (cf. pp. 29, 35)

$$\frac{d\delta s}{d\tau} = -2\delta p\,(1 + \mathrm{m} - \tfrac{5}{4}\mathrm{m}^2\cos 2\tau) - \tfrac{7}{2}\mathrm{m}^2\,\delta s\,\sin 2\tau$$

$$+ \frac{1}{V_1}\left[\int_0^\tau\left(\frac{\partial\Omega}{\partial x}\frac{dx}{d\tau} + \frac{\partial\Omega}{\partial y}\frac{dy}{d\tau} + \frac{\partial\Omega}{\partial z}\frac{dz}{d\tau}\right)d\tau + \tfrac{1}{2}\left\{\delta C - \frac{\kappa z^2}{r_1^3} - \left(\frac{dz}{d\tau}\right)^2\right\}\right]\dots\dots(44),$$

where δC will be chosen as is found most convenient. [In the previous work we chose $\delta C = 0$.]

By means of this equation we can eliminate δs from the differential equation for δp. For

$$2\frac{d\delta s}{d\tau}(1 + \mathrm{m} - \tfrac{5}{4}\mathrm{m}^2\cos 2\tau) + 7\mathrm{m}^2\,\delta s\,\sin 2\tau$$

$$= -4\delta p\,(1 + 2\mathrm{m} + \mathrm{m}^2 - \tfrac{5}{2}\mathrm{m}^2\cos 2\tau)$$

$$+ \frac{2}{V_1}(1 + \mathrm{m} - \tfrac{5}{4}\mathrm{m}^2\cos 2\tau)\left[\int_0^\tau\left(\frac{\partial\Omega}{\partial x}\frac{dx}{d\tau} + \frac{\partial\Omega}{\partial y}\frac{dy}{d\tau} + \frac{\partial\Omega}{\partial z}\frac{dz}{d\tau}\right)d\tau\right.$$

$$+ \left.\tfrac{1}{2}\left\{\delta C - \frac{\kappa z^2}{r_1^3} - \left(\frac{dz}{d\tau}\right)^2\right\}\right],$$

and therefore

$$\frac{d^2\delta p}{d\tau^2} + \delta p\,(1 + 2\mathrm{m} - \tfrac{1}{2}\mathrm{m}^2 - 15\mathrm{m}^2\cos 2\tau) = X\cos\phi + Y\sin\phi$$

$$+ \frac{2}{V_1}(1 + \mathrm{m} - \tfrac{5}{4}\mathrm{m}^2\cos 2\tau)\left[\int_0^\tau\left(\frac{\partial\Omega}{\partial x}\frac{dx}{d\tau} + \frac{\partial\Omega}{\partial y}\frac{dy}{d\tau} + \frac{\partial\Omega}{\partial z}\frac{dz}{d\tau}\right)d\tau\right.$$

$$+ \left.\tfrac{1}{2}\left\{\delta C - \frac{\kappa z^2}{r_1^3} - \left(\frac{dz}{d\tau}\right)^2\right\}\right]\dots\dots\dots\dots\dots(45).$$

We first neglect Ω, and consider X, Y as arising only from terms in z^2, i.e.

$$X = \frac{3}{2}\frac{\kappa z^2 x}{r_1^5}, \qquad Y = \frac{3}{2}\frac{\kappa z^2 y}{r_1^5}.$$

$$\therefore\ X\cos\phi + Y\sin\phi = \frac{3}{2}\frac{\kappa z^2}{r_1^5}(x\cos\phi + y\sin\phi).$$

To the required order of accuracy,

$$z = ka_0 \cos (g\tau + \zeta), \quad \frac{\kappa}{a_0{}^3} = 1 + 2m,$$

$$r_1 = a_0, \quad \phi = \tau, \quad x = a_0 \cos \tau, \quad y = a_0 \sin \tau.$$

$$\therefore \ X \cos \phi + Y \sin \phi = \tfrac{3}{4} (1 + 2m) \, k^2 a_0 \, [1 + \cos 2 \, (g\tau + \zeta)].$$

Also to order m

$$\frac{\kappa z^2}{r_1{}^3} + \left(\frac{dz}{d\tau}\right)^2 = (1 + 2m) \, k^2 a_0{}^2 \cos^2 (g\tau + \zeta) + g^2 \, k^2 a_0{}^2 \sin^2 (g\tau + \zeta)$$

$$= (1 + 2m) \, k^2 a_0{}^2,$$

since $g^2 = 1 + 2m$.

The equation for δp becomes therefore, as far as regards the new terms now introduced,

$$\frac{d^2 \delta p}{d\tau^2} + \delta p \, (1 + 2m) = \tfrac{3}{4} (1 + 2m) \, k^2 a_0 \, [1 + \cos 2 \, (g\tau + \zeta)]$$

$$+ \frac{(1 + m)}{a_0} \, [\delta C - (1 + 2m) \, k^2 a_0{}^2].$$

Hence $$\delta p - \tfrac{3}{4} k^2 a_0 - \frac{(1 - m)}{a_0} [\delta C - (1 + 2m) \, k^2 a_0{}^2]$$

$$= \tfrac{3}{4} \, \frac{1 + 2m}{1 + 2m - 4g^2} \, k^2 a_0 \cos 2 \, (g\tau + \zeta),^*$$

but $g^2 = 1 + 2m$, and therefore $1 + 2m - 4g^2 = - 3 \, (1 + 2m)$,

$$\therefore \ \delta p = \tfrac{3}{4} k^2 a_0 + \frac{(1 - m)}{a_0} [\delta C - (1 + 2m) \, k^2 a_0{}^2] - \tfrac{1}{4} k^2 a_0 \cos 2 \, (g\tau + \zeta).$$

Again the varied Jacobian integral is

$$\frac{d \delta s}{d\tau} = - 2 \, (1 + m) \, \delta p + \frac{1}{2a_0} [\delta C - (1 + 2m) \, k^2 a_0{}^2]$$

$$= - \tfrac{3}{2} (1 + m) \, k^2 a_0 - \frac{3}{2a_0} [\delta C - (1 + 2m) \, k^2 a_0{}^2] + \tfrac{1}{2} (1 + m) \, k^2 a_0 \cos 2 \, (g\tau + \zeta).$$

In order that δs may not increase with the time we choose δC so that the constant term is zero,

$$\therefore \ \delta C = m k^2 a_0,$$

and $$\frac{d \delta s}{d\tau} = \tfrac{1}{2} (1 + m) \, k^2 a_0 \cos 2 \, (g\tau + \zeta),$$

giving $$\delta s = \tfrac{1}{4} k^2 a_0 \sin 2 \, (g\tau + \zeta) \dots\dots\dots\dots\dots\dots\dots\dots(46),$$

as there is no need to introduce a new constant†. Using the value of δC just found we get

$$\delta p = - \tfrac{1}{4} k^2 a_0 - \tfrac{1}{4} k^2 a_0 \cos 2 \, (g\tau + \zeta) \ \dots\dots\dots\dots\dots(47).$$

Having obtained δp and δs, we now require δx, δy. These are

$$\delta x = \delta p \cos \phi - \delta s \sin \phi,$$

$$\delta y = \delta p \sin \phi + \delta s \cos \phi.$$

---

* It is of course only the special integral we require. The general integral when the right-hand side is zero has already been dealt with, § 7.

† Cf. same point in connection with equation (32).

In this case with sufficient accuracy $\phi = \tau$,

$$\delta x = - \tfrac{1}{4} a_0 k^2 \cos \tau - \tfrac{1}{4} a_0 k^2 \cos \tau \cos 2 (g\tau + \zeta) - \tfrac{1}{4} a_0 k^2 \sin \tau \sin 2 (g\tau + \zeta),$$

$$\delta y = - \tfrac{1}{4} a_0 k^2 \sin \tau - \tfrac{1}{4} a_0 k^2 \sin \tau \cos 2 (g\tau + \zeta) + \tfrac{1}{4} a_0 k^2 \cos \tau \sin 2 (g\tau + \zeta).$$

Dropping the recent use of X, Y in connection with the forces and using as before $X = x + \delta x$, $Y = y + \delta y$ we have

$$X = a_0 \cos \tau \, (1 - \tfrac{1}{4} k^2) - \tfrac{1}{4} a_0 k^2 \cos \tau \cos 2 (g\tau + \zeta) - \tfrac{1}{4} a_0 k^2 \sin \tau \sin 2 (g\tau + \zeta),$$

$$Y = a_0 \sin \tau \, (1 - \tfrac{1}{4} k^2) - \tfrac{1}{4} a_0 k^2 \sin \tau \cos 2 (g\tau + \zeta) + \tfrac{1}{4} a_0 k^2 \cos \tau \sin 2 (g\tau + \zeta),$$

$$R^2 = X^2 + Y^2 = a_0^2 \, (1 - \tfrac{1}{2} k^2) - \tfrac{1}{2} a_0^2 k^2 \cos 2 (g\tau + \zeta),$$

$$R = a_0 \, [1 - \tfrac{1}{4} k^2 - \tfrac{1}{4} k^2 \cos 2 (g\tau + \zeta)] \dots \dots \dots \dots \dots \dots \dots \dots \dots \dots \dots (48).$$

We thus get corrected result in radius vector as projected on to the ecliptic.

Again
$$\cos \Theta = \frac{X}{R} = \cos \tau - \tfrac{1}{4} k^2 \sin \tau \sin 2 (g\tau + \zeta),$$

$$\sin \Theta = \frac{Y}{R} = \sin \tau + \tfrac{1}{4} k^2 \cos \tau \sin 2 (g\tau + \zeta),$$

$$\Theta - \tau = \sin (\Theta - \tau) = \tfrac{1}{4} k^2 \sin 2 (g\tau + \zeta) \dots \dots \dots \dots \dots \dots (49).$$

Hence we have as a term in the moon's longitude $\tfrac{1}{4} k^2 \sin 2 (g\tau + \zeta)$. Terms of this type are called the reduction; they result from referring the moon's orbit to the ecliptic.

We have now only to consider the terms depending on Ω. We have seen that Ω vanishes when the solar eccentricity, e', is put equal to zero. We shall only develop Ω as far as first power of e'.

The radius vector r', and the true longitude v', of the sun are given to the required approximation by

$$r' = a' \, \{1 - e' \cos (n't - \varpi')\},$$

$$v' = n't + 2e' \sin (n't - \varpi').$$

Hence
$$x' = r' \cos (v' - n't) = r' = a' \, \{1 - e' \cos (n't - \varpi')\},$$

$$y' = r' \sin (v' - n't) \qquad = 2a'e' \sin (n't - \varpi').$$

And
$$n't = m\tau ;$$

$$\therefore \frac{xx' + yy'}{a'} = x - e'x \cos (m\tau - \varpi') + 2e'y \sin (m\tau - \varpi'),$$

$$\left(\frac{xx' + yy'}{a'} \right)^2 = x^2 - 2e'x^2 \cos (m\tau - \varpi') + 4e'xy \sin (m\tau - \varpi'),$$

$$\frac{a'^5}{r'^5} = 1 + 5e' \cos (m\tau - \varpi'),$$

$$\frac{3\mathrm{m}^2}{2}\left\{\frac{a'^3}{r'^5}(xx'+yy')^2 - x^2\right\} = \frac{9\mathrm{m}^2}{2}e'x^2\cos(\mathrm{m}\tau - \varpi') + 6\mathrm{m}^2e'xy\sin(\mathrm{m}\tau - \varpi'),$$

$$\tfrac{1}{2}\mathrm{m}^2(x^2+y^2+z^2)\left(1 - \frac{a'^3}{r'^3}\right) = -\tfrac{3}{2}\mathrm{m}^2(x^2+y^2+z^2)\,e'\cos(\mathrm{m}\tau - \varpi'),$$

$$\Omega = \mathrm{m}^2e'\left[3x^2\cos(\mathrm{m}\tau - \varpi') + 6xy\sin(\mathrm{m}\tau - \varpi') - \tfrac{3}{2}y^2\cos(\mathrm{m}\tau - \varpi')\right],$$

for we neglect m^2z^2 when multiplied by e',

$$\frac{\partial\Omega}{\partial x} = 6\mathrm{m}^2e'\left[x\cos(\mathrm{m}\tau - \varpi') + y\sin(\mathrm{m}\tau - \varpi')\right],$$

$$\frac{\partial\Omega}{\partial y} = 6\mathrm{m}^2e'\left[x\sin(\mathrm{m}\tau - \varpi') - \tfrac{1}{2}y\cos(\mathrm{m}\tau - \varpi')\right].$$

It is sufficiently accurate for us to take

$$x = a_0\cos\tau, \quad y = a_0\sin\tau,$$

$$\phi = \tau\,;$$

$$\therefore\ \frac{\partial\Omega}{\partial x}\cos\phi + \frac{\partial\Omega}{\partial y}\sin\phi = 6\mathrm{m}^2e'a_0\left[\cos^2\tau\cos(\mathrm{m}\tau - \varpi') + \cos\tau\sin\tau\sin(\mathrm{m}\tau - \varpi')\right.$$
$$\left. + \cos\tau\sin\tau\sin(\mathrm{m}\tau - \varpi') - \tfrac{1}{2}\sin^2\tau\cos(\mathrm{m}\tau - \varpi')\right]$$
$$= 3\mathrm{m}^2e'a_0\left[\cos(\mathrm{m}\tau - \varpi') + \cos 2\tau\cos(\mathrm{m}\tau - \varpi') + 2\sin 2\tau\sin(\mathrm{m}\tau - \varpi')\right]$$
$$- \tfrac{1}{2}\cos(\mathrm{m}\tau - \varpi') + \tfrac{1}{2}\cos 2\tau\cos(\mathrm{m}\tau - \varpi')$$
$$= 3\mathrm{m}^2e'a_0\left[\tfrac{1}{2}\cos(\mathrm{m}\tau - \varpi') + \tfrac{3}{4}\cos\{(2+\mathrm{m})\tau - \varpi'\} + \tfrac{3}{4}\cos\{(2-\mathrm{m})\tau + \varpi'\}\right.$$
$$\left. + \cos\{(2-\mathrm{m})\tau + \varpi'\} - \cos\{(2+\mathrm{m})\tau - \varpi'\}\right]$$
$$= \tfrac{3}{2}\mathrm{m}^2e'a_0\left[\cos(\mathrm{m}\tau - \varpi') - \tfrac{1}{2}\cos\{(2+\mathrm{m})\tau - \varpi'\} + \tfrac{7}{2}\cos\{(2-\mathrm{m})\tau + \varpi'\}\right].$$

Again

$$\frac{\partial\Omega}{\partial x}\frac{dx}{d\tau} + \frac{\partial\Omega}{\partial y}\frac{dy}{d\tau} = 6\mathrm{m}^2e'a_0\left[-\sin\tau\cos\tau\cos(\mathrm{m}\tau - \varpi') - \sin^2\tau\sin(\mathrm{m}\tau - \varpi')\right.$$
$$\left. + \cos^2\tau\sin(\mathrm{m}\tau - \varpi') - \tfrac{1}{2}\sin\tau\cos\tau\cos(\mathrm{m}\tau - \varpi')\right]$$
$$= 3\mathrm{m}^2e'a_0\left[-\tfrac{3}{2}\sin 2\tau\cos(\mathrm{m}\tau - \varpi') + 2\cos 2\tau\sin(\mathrm{m}\tau - \varpi')\right]$$
$$= \tfrac{3}{2}\mathrm{m}^2e'a_0\left[-\tfrac{3}{2}\sin\{(2+\mathrm{m})\tau - \varpi'\} - \tfrac{3}{2}\sin\{(2-\mathrm{m})\tau + \varpi'\}\right.$$
$$\left. + 2\sin\{(2+\mathrm{m})\tau - \varpi'\} - 2\sin\{(2-\mathrm{m})\tau + \varpi'\}\right]$$
$$= \tfrac{3}{2}\mathrm{m}^2e'a_0\left[\tfrac{1}{2}\sin\{(2+\mathrm{m})\tau - \varpi'\} - \tfrac{7}{2}\sin\{(2-\mathrm{m})\tau + \varpi'\}\right],$$

$$2\int\left(\frac{\partial\Omega}{\partial x}\frac{dx}{d\tau} + \frac{\partial\Omega}{\partial y}\frac{dy}{d\tau}\right)d\tau = -\tfrac{3}{2}\mathrm{m}^2e'a_0\left[\tfrac{1}{2}\cos\{(2+\mathrm{m})\tau - \varpi'\}\right.$$
$$\left. - \tfrac{7}{2}\cos\{(2-\mathrm{m})\tau + \varpi'\}\right];$$

$$\therefore\ \frac{\partial\Omega}{\partial x}\cos\phi + \frac{\partial\Omega}{\partial y}\sin\phi + 2\int\left(\frac{\partial\Omega}{\partial x}\frac{dx}{d\tau} + \frac{\partial\Omega}{\partial y}\frac{dy}{d\tau}\right)d\tau = \tfrac{3}{2}\mathrm{m}^2e'a_0\left[\cos(\mathrm{m}\tau - \varpi')\right.$$
$$\left. - \cos\{(2+\mathrm{m})\tau - \varpi'\} + 7\cos\{(2-\mathrm{m})\tau + \varpi'\}\right].$$

Hence to the order required

$$\frac{d^2\delta p}{d\tau^2} + (1 + 2\mathrm{m})\,\delta p = \tfrac{3}{2}\mathrm{m}^2 e' a_0 \left[\cos(\mathrm{m}\tau - \varpi') - \cos\{(2 + \mathrm{m})\,\tau - \varpi'\}\right.$$
$$\left. + 7\cos\{(2 - \mathrm{m})\,\tau + \varpi'\}\right],$$

$$\delta p = \tfrac{3}{2}\mathrm{m}^2 e' a_0 \left[\frac{\cos(\mathrm{m}\tau - \varpi')}{-\mathrm{m}^2 + 1 + 2\mathrm{m}} - \frac{\cos\{(2 + \mathrm{m})\,\tau - \varpi'\}}{-(4 + 4\mathrm{m}) + 1 + 2\mathrm{m}} + \frac{7\cos\{(2 - \mathrm{m})\,\tau + \varpi'\}}{-(4 - 4\mathrm{m}) + 1 + 2\mathrm{m}}\right]$$

$$= \tfrac{3}{2}\mathrm{m}^2 e' a_0 \left[\cos(\mathrm{m}\tau - \varpi') + \tfrac{1}{3}\cos\{(2 + \mathrm{m})\,\tau - \varpi'\} - \tfrac{7}{3}\cos\{(2 - \mathrm{m})\,\tau + \varpi'\}\right]$$
$$\dots\dots(50),$$

$$\frac{d\delta s}{d\tau} = -2\delta p\,(1 + \mathrm{m}) + \frac{1}{V}\int\left(\frac{\partial\Omega}{\partial x}\frac{dx}{d\tau} + \frac{\partial\Omega}{\partial y}\frac{dy}{d\tau}\right) d\tau$$

$$= -3\mathrm{m}^2 e' a_0\left[\cos(\mathrm{m}\tau - \varpi') + \tfrac{1}{3}\cos\{(2 + \mathrm{m})\,\tau - \varpi'\} - \tfrac{7}{3}\cos\{(2 - \mathrm{m})\,\tau + \varpi'\}\right]$$
$$- \tfrac{3}{4}\mathrm{m}^2 e'\left[\tfrac{1}{2}\cos\{(2 + m)\,\tau - \varpi'\} - \tfrac{7}{2}\cos\{(2 - \mathrm{m})\,\tau + \varpi'\}\right]$$

$$= -3\mathrm{m}^2 e' a_0\left[\cos(\mathrm{m}\tau - \varpi') + \tfrac{11}{24}\cos\{(2 + \mathrm{m})\,\tau - \varpi'\}\right.$$
$$\left. - \tfrac{77}{24}\cos\{(2 - \mathrm{m})\,\tau + \varpi'\}\right];$$

$$\therefore\ \delta s = -3\mathrm{m}e' a_0\sin(\mathrm{m}\tau - \varpi') - 3\mathrm{m}^2 e' a_0\left[\tfrac{11}{48}\sin\{(2 + \mathrm{m})\,\tau - \varpi'\}\right.$$
$$\left. - \tfrac{77}{48}\sin\{(2 - \mathrm{m})\,\tau + \varpi'\}\right]\dots\dots(51).$$

Hence to order $\mathrm{m}e'$, to which order only our result is correct,

$$\delta p = 0, \quad \delta s = -3\mathrm{m}e' a_0\sin(\mathrm{m}\tau - \varpi').$$

And following our usual method for obtaining new terms in radius vector and longitude

$$\delta x = \delta p\cos\phi - \delta s\sin\phi, \quad \delta y = \delta p\sin\phi + \delta s\cos\phi,$$
$$\delta x = -\delta s\sin\tau, \qquad\qquad \delta y = \delta s\cos\tau,$$
$$X = a_0\left[\cos\tau + 3\mathrm{m}e'\sin\tau\sin(\mathrm{m}\tau - \varpi')\right],$$
$$Y = a_0\left[\sin\tau - 3\mathrm{m}e'\cos\tau\sin(\mathrm{m}\tau - \varpi')\right],$$
$$R^2 = a_0^2\left[1 + 3\mathrm{m}e'\sin 2\tau\sin(\mathrm{m}\tau - \varpi') - 3\mathrm{m}e'\sin 2\tau\sin(\mathrm{m}\tau - \varpi')\right] = a_0^2$$
$$\dots\dots(52),$$

and to the order required there is no term in radius vector

$$\cos\Theta = \cos\tau + 3\mathrm{m}e'\sin\tau\sin(\mathrm{m}\tau - \varpi'),$$
$$\sin\Theta = \sin\tau - 3\mathrm{m}e'\cos\tau\sin(\mathrm{m}\tau - \varpi'),$$
$$\sin(\Theta - \tau) = -3\mathrm{m}e'\sin(\mathrm{m}\tau - \varpi'),$$
$$\Theta = \tau - 3\mathrm{m}e'\sin(\mathrm{m}\tau - \varpi')\dots\dots\dots\dots\dots\dots(53).$$

The new term in the longitude is $-3\mathrm{m}e'\sin(l' - \varpi')$. This term is called the annual equation.

§ 11. *Compilation of Results.*

Let v be the longitude, s the tangent of the latitude (or to our order simply the latitude).   When we collect our results we find

$$v = \quad\underset{\substack{\text{(mean} \\ \text{longitude} \\ =nt+\epsilon)}}{l} \quad \underset{\substack{\text{equation to} \\ \text{the centre}}}{+\, 2e \sin (l - \varpi)} + \underset{\text{evection}}{\tfrac{15}{4}me \sin (l - 2l' + \varpi)} + \underset{\text{variation}}{\tfrac{11}{8}m^2 \sin 2 (l - l')}$$

$$\underset{\text{reduction}}{-\, \tfrac{1}{4}k^2 \sin 2 (l - \Omega)} - \underset{\text{annual equation}}{3me' \sin (l' - \varpi')},$$

$$s = k \sin (l - \Omega) + \underset{\text{evection in latitude}}{\tfrac{3}{8}mk \sin (l - 2l' + \Omega)}.$$

For R, the projection of the radius vector on the ecliptic, we get

$$R = a\,[\,1 - \tfrac{1}{6}m^2 - \tfrac{1}{4}k^2 - \underset{\substack{\text{equation to the} \\ \text{centre}}}{e \cos (l - \varpi)} - \underset{\text{evection}}{\tfrac{15}{8}me \cos (l - 2l' + \varpi)} - \underset{\text{variation}}{m^2 \cos 2 (l - l')}$$

$$+ \underset{\text{reduction}}{\tfrac{1}{4}k^2 \cos 2 (l - \Omega)]} \dots\dots(54).$$

To get the actual radius vector we require to multiply by sec β, i.e. by

$$1 + \tfrac{1}{2}k^2 \sin^2 (l - \Omega) \ \ \text{or} \ \ 1 + \tfrac{1}{4}k^2 - \tfrac{1}{4}k^2 \cos 2 (l - \Omega).$$

This amounts to removing the terms $-\tfrac{1}{4}k^2 + \tfrac{1}{4}k^2 \cos 2 (l - \Omega)$.   The radius vector then is

$$a\,[\,1 - \tfrac{1}{6}m^2 - e \cos (l - \varpi) - \tfrac{15}{8}me \cos (l - 2l' + \varpi) - m^2 \cos 2 (l - l')].$$

This is independent of k, but k will enter into product terms of higher order than we have considered.   The perturbations are excluded by putting $m = 0$ and the value of the radius vector is then independent of k as it should be.   The quantity of practical importance is not the radius vector but its reciprocal.   To our degree of approximation it is

$$\frac{1}{a}\,[\,1 + \tfrac{1}{6}m^2 + e \cos (l - \varpi) + \tfrac{15}{8}me \cos (l - 2l' + \varpi) + m^2 \cos 2 (l - l')].$$

It may be noted in conclusion that the terms involving only e in the coefficient, and designated the equation to the centre, are not perturbations but the ordinary elliptic inequalities.   There are terms in e^2 but these have not been included in our work.

NOTE 1. *On the Infinite Determinant of* § 5.

We assume (as has been justified by Poincaré) that we may treat the infinite determinant as though it were a finite one.

For every row corresponding to $+i$ there is another corresponding to $-i$, and there is one for $i = 0$.

If we write $-c$ for c the determinant is simply turned upside down. Hence the roots occur in pairs and if c_0 is a root $-c_0$ is also a root.

If for c we write $c \pm 2j$, where j is an integer, we simply shift the centre of the determinant.

Hence if c_0 is a root, $\pm c_0 \pm 2j$ are also roots.

But these are the roots of $\cos \pi c = \cos \pi c_0$.

Therefore the determinant must be equal to

$$k \, (\cos \pi c - \cos \pi c_0).$$

If all the roots have been enumerated, k is independent of c.

Now the number of roots cannot be affected by the values assigned to the Θ's. Let us put $\Theta_1 = \Theta_2 = \Theta_3 = \ldots = 0$.

The determinant then becomes equal to the product of the diagonal terms and the equation is

$$\ldots [\Theta_0 - (c - 2)^2] \, [\Theta_0 - c^2] \, [\Theta_0 - (c + 2)^2] \ldots = 0.$$

$c_0 = \pm \sqrt{\Theta_0}$ is one pair of roots, and all the others are given by $c_0 \pm 2i$.

Hence there are no more roots and k is independent of c.

The determinant which we have obtained is inconvenient because the diagonal elements increase as we pass away from the centre while the non-diagonal elements are of the same order of magnitude for all the rows. But the roots of the determinant are not affected if the rows are multiplied by numerical constants and we can therefore introduce such numerical multipliers as we may find convenient.

The following considerations indicate what multipliers may prove useful. If we take a finite determinant from the centre of the infinite one it can be completely expanded by the ordinary processes. Each of the terms in the expansion will only involve c through elements from the principal diagonal and the term obtained by multiplying all the elements of this diagonal will contain the highest power of c. When the determinant has $(2i + 1)$ rows and columns, the highest power of c will be $-c^{2(2i+1)}$. We wish to associate the infinite determinant with $\cos \pi c$. Now

$$\cos \pi c = \left(1 - \frac{4c^2}{1}\right)\left(1 - \frac{4c^2}{9}\right)\left(1 - \frac{4c^2}{25}\right)\ldots.$$

The first $2i + 1$ terms of this product may be written

$$\left(1 - \frac{2c}{4i+1}\right)\left(1 - \frac{2c}{4i-1}\right)\cdots\left(1 + \frac{2c}{4i-1}\right)\left(1 + \frac{2c}{4i+1}\right),$$

and the highest power of c in this product is

$$\frac{4c^2}{(4i)^2 - 1} \cdot \frac{4c^2}{\{4\,(i-1)\}^2 - 1} \cdots \frac{4c^2}{(4i)^2 - 1}.$$

Hence we multiply the ith row below or above the central row by $\dfrac{-4}{(4i)^2 - 1}$.

The ith diagonal term below the central term will now be $\dfrac{4\,[(2i + c)^2 - \Theta_0]}{(4i)^2 - 1}$ and will be denoted by $\{i\}$. It clearly tends to unity as i tends to infinity by positive or negative values. The ith row below the central row will now read

$$\cdots \frac{-4\Theta_2}{(4i)^2 - 1},\quad \frac{-4\Theta_1}{(4i)^2 - 1},\quad \{i\},\quad \frac{-4\Theta_1}{(4i)^2 - 1},\quad \frac{-4\Theta_2}{(4i)^2 - 1},\quad \cdots.$$

The new determinant which we will denote by $\nabla\,(c)$ has the same roots as the original one and so we may write

$$\nabla\,(c) = k'\,\{\cos \pi c - \cos \pi c_0\},$$

where k' is a new numerical constant. But it is easy to see that $k' = 1$. This was the object of introducing the multipliers and that it is true is easily proved by taking the case of $\Theta_1 = \Theta_2 = \ldots = 0$ and $\Theta_0 = \frac{1}{4}$, in which case the determinant reduces to $\cos \pi c$. We thus have the equation

$$\nabla\,(c) = \cos \pi c - \cos \pi c_0,$$

which can be considered as an identity in c.

Putting $c = 0$ we get

$$\nabla\,(0) = 1 - \cos \pi c_0.$$

$\nabla\,(0)$ depends only on the Θ's; written so as to shew the principal elements it is

$$\begin{vmatrix}
\cdots & \frac{4}{63}(16 - \Theta_0), & -\frac{4}{63}\Theta_1, & -\frac{4}{63}\Theta_2, & -\frac{4}{63}\Theta_3, & -\frac{4}{63}\Theta_4, & \cdots \\
\cdots & -\frac{4}{15}\Theta_1, & \frac{4}{15}(4 - \Theta_0), & -\frac{4}{15}\Theta_1, & -\frac{4}{15}\Theta_2, & -\frac{4}{15}\Theta_3, & \cdots \\
\cdots & 4\,\Theta_2, & 4\,\Theta_1, & 4\,\Theta_0, & 4\,\Theta_1, & 4\,\Theta_2, & \cdots \\
\cdots & -\frac{4}{15}\Theta_3, & -\frac{4}{15}\Theta_2, & -\frac{4}{15}\Theta_1, & \frac{4}{15}(4 - \Theta_0), & -\frac{4}{15}\Theta_1, & \cdots \\
\cdots & -\frac{4}{63}\Theta_4, & -\frac{4}{63}\Theta_3, & -\frac{4}{63}\Theta_2, & -\frac{4}{63}\Theta_1, & \frac{4}{63}(16 - \Theta_0), & \cdots
\end{vmatrix}$$

If Θ_1, Θ_2, etc. vanish, the solution of the differential equation is $\cos(\sqrt{\Theta_0} + \epsilon)$ or $c = \sqrt{\Theta_0}$. But in this case the determinant has only diagonal terms and the product of the diagonal terms of $\nabla\,(0)$ is $1 - \cos \pi \sqrt{\Theta_0}$ or $2 \sin^2 \frac{1}{2}\pi \sqrt{\Theta_0}$.

Hence we may divide each row by its diagonal member and put $2 \sin^2 \tfrac{1}{2}\pi \sqrt{\Theta_0}$ outside.

If therefore

$$\Delta\,(0)= \begin{vmatrix} \cdots & 1 & , & -\dfrac{\Theta_1}{16-\Theta_0}, & -\dfrac{\Theta_2}{16-\Theta_0}, & -\dfrac{\Theta_3}{16-\Theta_0}, & -\dfrac{\Theta_4}{16-\Theta_0}, & \cdots \\[2ex] \cdots & -\dfrac{\Theta_1}{4-\Theta_0}, & 1 & , & -\dfrac{\Theta_1}{4-\Theta_0}, & -\dfrac{\Theta_2}{4-\Theta_0}, & -\dfrac{\Theta_3}{4-\Theta_0}, & \cdots \\[2ex] \cdots & \dfrac{\Theta_2}{\Theta_0}, & \dfrac{\Theta_1}{\Theta_0}, & 1 & , & \dfrac{\Theta_1}{\Theta_0}, & \dfrac{\Theta_2}{\Theta_0}, & \cdots \\[2ex] \cdots & -\dfrac{\Theta_3}{4-\Theta_0}, & -\dfrac{\Theta_2}{4-\Theta_0}, & -\dfrac{\Theta_1}{4-\Theta_0}, & 1 & , & -\dfrac{\Theta_1}{4-\Theta_0}, & \cdots \end{vmatrix}$$

$\nabla\,(0) = 2 \sin^2 \tfrac{1}{2}\pi \sqrt{\Theta_0}\, \Delta\,(0).$

Now since $\cos \pi c_0 = 1 - \nabla\,(0) = 1 - 2 \sin^2 \tfrac{1}{2}\pi \sqrt{\Theta_0}\, \Delta\,(0),$

we have $\dfrac{\sin^2 \tfrac{1}{2}\pi c_0}{\sin^2 \tfrac{1}{2}\pi \sqrt{\Theta_0}} = \Delta\,(0),$

an equation to be solved for c_0 (or c).

Clearly for stability $\Delta\,(0)$ must be positive and $\Delta\,(0) < \operatorname{cosec}^2 \tfrac{1}{2}\pi \sqrt{\Theta_0}.$ Hill gives other transformations.

NOTE 2*.   *On the periodicity of the integrals of the equation*

$$\frac{d^2\delta p}{d\tau^2} + \Theta\delta p = 0,$$

where $\Theta = \Theta_0 + \Theta_1 \cos 2\tau + \Theta_2 \cos 4\tau + \dots.$

Since the equation remains unchanged when τ becomes $\tau + \pi$, it follows that if $\delta p = F(\tau)$ is a solution $F(\tau+\pi)$ is also a solution.

Let $\phi(\tau)$ be a solution subject to the conditions that when

$$\tau = 0, \quad \delta p = 1, \quad \frac{d\delta p}{d\tau} = 0 \,; \text{ i.e. } \phi(0) = 1, \quad \phi'(0) = 0.$$

Let $\psi(\tau)$ be a second solution subject to the conditions that when

$$\tau = 0, \quad \delta p = 0, \quad \frac{d\delta p}{d\tau} = 1 \,; \text{ i.e. } \psi(0) = 0, \quad \psi'(0) = 1.$$

---

* This treatment of the subject was pointed out to Sir George Darwin by Mr S. S. Hough.

It is clear that $\phi(\tau)$ is an even function of τ, and $\psi(\tau)$ an odd one, so that

$$\phi(-\tau) = \phi(\tau), \qquad \psi(-\tau) = -\psi(\tau),$$
$$\phi'(-\tau) = -\phi'(\tau), \qquad \psi'(-\tau) = \psi'(\tau).$$

Then the general solution of the equation is

$$\delta p = F(\tau) = A\phi(\tau) + B\psi(\tau),$$

where A and B are two arbitrary constants.

Since $\phi(\tau + \pi)$, $\psi(\tau + \pi)$ are also solutions of the equation, it follows that

$$\left. \begin{array}{l} \phi(\tau + \pi) = \alpha\phi(\tau) + \beta\psi(\tau) \\ \psi(\tau + \pi) = \gamma\phi(\tau) + \delta\psi(\tau) \end{array} \right\} \quad \dots\dots\dots\dots\dots(55),$$

where α, β, γ, δ are definite constants.

If possible let $A : B$ be so chosen that

$$F(\tau + \pi) = \nu F(\tau),$$

where ν is a numerical constant.

When we substitute for F its values in terms of ϕ and ψ, we obtain

$$A\phi(\tau + \pi) + B\psi(\tau + \pi) = \nu [A\phi(\tau) + B\psi(\tau)].$$

Further, substituting for $\phi(\tau + \pi)$, $\psi(\tau + \pi)$ their values, we have

$$A[\alpha\phi(\tau) + \beta\psi(\tau)] + B[\gamma\phi(\tau) + \delta\psi(\tau)] = \nu [A\phi(\tau) + B\psi(\tau)],$$

whence $\quad [A(\alpha - \nu) + B\gamma]\phi(\tau) + [A\beta + B(\delta - \nu)]\psi(\tau) = 0.$

Since this is satisfied for all values of τ,

$$A(\alpha - \nu) + B\gamma = 0,$$
$$A\beta + B(\delta - \nu) = 0,$$
$$\therefore \quad (\alpha - \nu)(\delta - \nu) - \beta\gamma = 0,$$
$$\text{i.e.} \quad \nu^2 - (\alpha + \delta)\nu + \alpha\delta - \beta\gamma = 0,$$

an equation for ν in terms of the constants α, β, γ, δ. This equation can be simplified.

Since $\qquad \dfrac{d^2\phi}{d\tau^2} + \Theta\phi = 0, \qquad \dfrac{d^2\psi}{d\tau^2} + \Theta\psi = 0,$

we have $\qquad \phi\dfrac{d^2\psi}{d\tau^2} - \psi\dfrac{d^2\phi}{d\tau^2} = 0.$

On integration of which

$$\phi\psi' - \psi\phi' = \text{const.}$$

But $\qquad \phi(0) = 1, \quad \psi'(0) = 1, \quad \psi(0) = 0, \quad \phi'(0) = 0.$

Therefore the constant is unity; and

$$\phi(\tau)\psi'(\tau) - \psi(\tau)\phi'(\tau) = 1 \quad \dots\dots\dots\dots\dots(56).$$

But putting $\tau = 0$ in the equations (55), and in the equations obtained by differentiating them,

$$\phi(\pi) = \alpha\phi(0) + \beta\psi(0) = \alpha,$$

$$\psi(\pi) = \gamma\phi(0) + \delta\psi(0) = \gamma,$$

$$\phi'(\pi) = \alpha\phi'(0) + \beta\psi'(0) = \beta,$$

$$\psi'(\pi) = \gamma\phi'(0) + \delta\psi'(0) = \delta.$$

Therefore by (56), $\alpha\delta - \beta\gamma = 1.$

Accordingly our equation for ν is

$$\nu^2 - (\alpha + \delta)\nu + 1 = 0$$

or

$$\tfrac{1}{2}\left(\nu + \frac{1}{\nu}\right) = \tfrac{1}{2}(\alpha + \delta).$$

If now we put $\tau = -\tfrac{1}{2}\pi$ in (55) and the equations obtained by differentiating them,

$$\phi(\tfrac{1}{2}\pi) = \alpha\phi(-\tfrac{1}{2}\pi) + \beta\psi(-\tfrac{1}{2}\pi) = \alpha\phi(\tfrac{1}{2}\pi) - \beta\psi(\tfrac{1}{2}\pi),$$

$$\psi(\tfrac{1}{2}\pi) = \gamma\phi(-\tfrac{1}{2}\pi) + \delta\psi(-\tfrac{1}{2}\pi) = \gamma\phi(\tfrac{1}{2}\pi) - \delta\psi(\tfrac{1}{2}\pi),$$

$$\phi'(\tfrac{1}{2}\pi) = \alpha\phi'(-\tfrac{1}{2}\pi) + \beta\psi'(-\tfrac{1}{2}\pi) = -\alpha\phi'(\tfrac{1}{2}\pi) + \beta\psi'(\tfrac{1}{2}\pi),$$

$$\psi'(\tfrac{1}{2}\pi) = \gamma\phi'(-\tfrac{1}{2}\pi) + \delta\psi'(-\tfrac{1}{2}\pi) = -\gamma\phi'(\tfrac{1}{2}\pi) + \delta\psi'(\tfrac{1}{2}\pi),$$

$$\frac{\phi(\tfrac{1}{2}\pi)}{\psi(\tfrac{1}{2}\pi)} = \frac{\beta}{\alpha - 1} = \frac{\delta + 1}{\gamma}, \quad \frac{\psi'(\tfrac{1}{2}\pi)}{\phi'(\tfrac{1}{2}\pi)} = \frac{\alpha + 1}{\beta} = \frac{\gamma}{\delta - 1},$$

$$\frac{\phi(\tfrac{1}{2}\pi)\,\psi'(\tfrac{1}{2}\pi)}{\psi(\tfrac{1}{2}\pi)\,\phi'(\tfrac{1}{2}\pi)} = \frac{\alpha + 1}{\alpha - 1} = \frac{\delta + 1}{\delta - 1}.$$

But since $\phi(\tfrac{1}{2}\pi)\,\psi'(\tfrac{1}{2}\pi) - \phi'(\tfrac{1}{2}\pi)\,\psi(\tfrac{1}{2}\pi) = 1$ we have

$$\alpha = \delta = \tfrac{1}{2}(\alpha + \delta) = \phi(\tfrac{1}{2}\pi)\,\psi'(\tfrac{1}{2}\pi) + \phi'(\tfrac{1}{2}\pi)\,\psi(\tfrac{1}{2}\pi).$$

Hence the equation for ν may be written in five different forms, viz.

$$\tfrac{1}{2}\left(\nu + \frac{1}{\nu}\right) = \phi(\pi) = \psi'(\pi) = \phi(\tfrac{1}{2}\pi)\,\psi'(\tfrac{1}{2}\pi) + \phi'(\tfrac{1}{2}\pi)\,\psi(\tfrac{1}{2}\pi)$$

$$= 1 + 2\phi'(\tfrac{1}{2}\pi)\,\psi(\tfrac{1}{2}\pi) = 2\phi(\tfrac{1}{2}\pi)\,\psi'(\tfrac{1}{2}\pi) - 1 \quad \ldots\ldots\ldots\ldots(57).$$

It remains to determine the meaning of ν in terms of the c introduced in the solution by means of the infinite determinant.

The former solution was

$$\delta p = \sum_{-\infty}^{+\infty}\{A_j \cos(c + 2j)\tau + B_j \sin(c + 2j)\tau\},$$

where $A_j : B_j$ as $-\cos\epsilon : \sin\epsilon.$

In the solution $\phi(\tau)$ we have $\phi(0) = 1$, $\phi'(0) = 0$, and $\phi(\tau)$ is an even function of τ. Hence to get $\phi(\tau)$ from δp we require to put $\Sigma A_j = 1$, and $B_j = 0$ for all values of j.

This gives
$$\phi(\pi) = \Sigma \left\{ A_j \cos(c + 2j)\pi \right\}$$
$$= \cos \pi c \, \Sigma A_j = \cos \pi c.$$

Similarly we may shew that $\psi'(\pi) = \cos \pi c$.

It follows from equations (57) that
$$\cos \pi c = \phi(\pi) = \psi'(\pi),$$
$$\cos^2 \tfrac{1}{2}\pi c = \phi(\tfrac{1}{2}\pi)\,\psi'(\tfrac{1}{2}\pi); \quad \sin^2 \tfrac{1}{2}\pi c = -\,\phi'(\tfrac{1}{2}\pi)\,\psi(\tfrac{1}{2}\pi).$$

We found on p. 55 that $\sin^2 \tfrac{1}{2}\pi c = \sin^2 \tfrac{1}{2}\pi \sqrt{\Theta_0} \,.\, \Delta(0)$, where $\Delta(0)$ is a certain determinant. Hence the last solution being of this form, we have the value of the determinant $\Delta(0)$ in terms of ϕ and ψ, viz.

$$\Delta(0) = -\,\frac{\phi'(\tfrac{1}{2}\pi)\,\psi(\tfrac{1}{2}\pi)}{\sin^2 \tfrac{1}{2}\pi \sqrt{\Theta_0}}\,.$$

From this new way of looking at the matter it appears that the value of c may be found by means of the two special solutions ϕ and ψ.

ON LIBRATING PLANETS AND ON A NEW FAMILY OF PERIODIC ORBITS

§ 1. *Librating Planets.*

In Professor Ernest Brown's interesting paper on "A New Family of Periodic Orbits" (*M.N.*, *R.A.S.*, vol. LXXI., 1911, p. 438) he shews how to obtain the orbit of a planet which makes large oscillations about the vertex of the Lagrangian equilateral triangle. In discussing this paper I shall depart slightly from his notation, and use that of my own paper on "Periodic Orbits" (*Scientific Papers*, vol. IV., or *Acta Math.*, vol. LI.). "Jove," J, of mass 1, revolves at distance 1 about the "Sun," S, of mass ν, and the orbital angular velocity is n, where $n^2 = \nu + 1$.

The axes of reference revolve with SJ as axis of x, and the heliocentric and jovicentric rectangular coordinates of the third body are x, y and $x - 1$, y respectively. The heliocentric and jovicentric polar co-ordinates are respectively r, θ and ρ, ψ. The potential function for relative energy is Ω.

The equations of motion and Jacobian integral, from which Brown proceeds, are

$$\left.\begin{aligned}
\frac{d^2 r}{dt^2} - r\frac{d\theta}{dt}\left(\frac{d\theta}{dt} + 2n\right) &= \frac{\partial\Omega}{\partial r} \\
\frac{d}{dt}\left[r^2\left(\frac{d\theta}{dt} + n\right)\right] &= \frac{\partial\Omega}{\partial\theta} \\
\left(\frac{dr}{dt}\right)^2 + \left(r\frac{d\theta}{dt}\right)^2 &= 2\Omega - C \\
2\Omega = \nu\left(r^2 + \frac{2}{r}\right) + \left(\rho^2 + \frac{2}{\rho}\right)
\end{aligned}\right\} \quad\ldots\ldots\ldots\ldots\ldots\ldots(1).$$

where

The following are rigorous transformations derived from these equations, virtually given by Brown in approximate forms in equation (13), and at the foot of p. 443 :—

$$\left(\frac{d\theta}{dt} + n\right)^2 = A + \frac{1}{r}\frac{d^2r}{dt^2} \quad\dots\dots\dots\dots\dots\dots\dots\dots\dots\dots\dots\dots\dots(2),$$

$$\frac{dr}{dt}\left(L + 3\frac{d^2r}{dt^2}\right) = B + D\left(\frac{d\theta}{dt} + n\right) - r\frac{d^3r}{dt^3} \quad\dots\dots\dots\dots\dots\dots\dots\dots(3),$$

$$\frac{d^2r}{dt^2}\left(L + 3\frac{d^2r}{dt^2}\right) = E\left(\frac{dr}{dt}\right)^2 + F\frac{dr}{dt}\frac{d\theta}{dt} + G\left(\frac{d\theta}{dt}\right)^2 + H\frac{dr}{dt} + J\frac{d\theta}{dt} + K$$

$$- 4\frac{dr}{dt}\frac{d^3r}{dt^3} - r\frac{d^4r}{dt^4}\dots\dots(4),$$

where

$$A = n^2 - \frac{\partial\Omega}{r\partial r} = \frac{\nu}{r^3} + 1 - \frac{1}{r}\left(\rho - \frac{1}{\rho^2}\right)\cos(\theta - \psi),$$

$$B = -nr\frac{\partial^2\Omega}{\partial r\partial\theta} = -n\sin\psi\left[\left(\rho - \frac{1}{\rho^2}\right) + \frac{3r}{\rho^3}\cos(\theta - \psi)\right],$$

$$D = r\frac{\partial^2\Omega}{\partial r\partial\theta} + 2\frac{\partial\Omega}{\partial\theta} = 3\sin\psi\left[\left(\rho - \frac{1}{\rho^2}\right) + \frac{r}{\rho^3}\cos(\theta - \psi)\right],$$

$$L = 4n^2r - r\frac{\partial^2\Omega}{\partial r^2} - 3\frac{\partial\Omega}{\partial r} = \frac{\nu}{r^2} + 3r + \frac{r}{\rho^3} - 3\left(\rho - \frac{1}{\rho^2}\right)\cos(\theta - \psi) - \frac{3r}{\rho^3}\cos^2(\theta - \psi),$$

$$E = r\frac{\partial^3\Omega}{\partial r^3} + 4\frac{\partial^2\Omega}{\partial r^2} - 4n^2 = \frac{2\nu}{r^3} + \frac{4}{\rho^3}[3\cos^2(\theta - \psi) - 1]$$

$$+ \frac{3r}{\rho^4}\cos(\theta - \psi)[3 - 5\cos^2(\theta - \psi)],$$

$$F = 2r\frac{\partial^3\Omega}{\partial r^2\partial\theta} + 4\frac{\partial^2\Omega}{\partial r\partial\theta} - 4\frac{\partial\Omega}{r\partial\theta} = \frac{6}{\rho^4}\sin\psi[5r\sin^2(\theta - \psi) - 4\cos\theta],$$

$$G = r\frac{\partial^3\Omega}{\partial r\partial\theta^2} + 2\frac{\partial^2\Omega}{\partial\theta^2} = \frac{3r}{\rho}\left[\left(\rho - \frac{1}{\rho^2}\right)\cos\theta - \frac{r}{\rho^3}\sin\psi(5\sin^2(\theta - \psi) - 1)\right],$$

$$H = -\frac{4n}{r}\frac{\partial\Omega}{\partial\theta} = 4n\left(\rho - \frac{1}{\rho^2}\right)\sin(\theta - \psi),$$

$$J = 2n\frac{\partial^2\Omega}{\partial\theta^2} = \frac{2nr}{\rho}\left[\left(\rho - \frac{1}{\rho^2}\right)\cos\theta - \frac{3}{\rho^2}\sin\psi\sin(\theta - \psi)\right],$$

$$K = \frac{\partial\Omega}{r^2\partial\theta}\left(r\frac{\partial^2\Omega}{\partial r\partial\theta} + 2\frac{\partial\Omega}{\partial\theta}\right) = \frac{3}{r}\left(\rho - \frac{1}{\rho^2}\right)\sin\theta\sin\psi\left(1 + \frac{1}{\rho^4}\cos\psi\right).$$

A great diversity of forms might be given to these functions, but the foregoing seemed to be as convenient for computation as I could devise.

It is known that when ν is less than $24\cdot9599$* the vertex of the equilateral triangle is an unstable solution of the problem, and if the body is displaced from the vertex it will move away in a spiral orbit. Hence for small values of ν there are no small closed periodic orbits of the kind considered by Brown. But certain considerations led him to conjecture that

* " Periodic Orbits," *Scientific Papers*, vol. IV., p. 73.

there might still exist large oscillations of this kind. The verification of such a conjecture would be interesting, and in my attempt to test his idea I took ν equal to 10. This value was chosen because the results will thus form a contribution towards that survey of periodic orbits which I have made in previous papers for ν equal to 10.

Brown's system of approximation, which he justifies for large values of ν, may be described, as far as it is material for my present object, as follows:—

We begin the operation at any given point r, θ, such that ρ is greater than unity.

Then in (2) and (3) $\dfrac{d^2r}{dt^2}$ and $\dfrac{d^3r}{dt^3}$ are neglected, and we thence find $\dfrac{dr}{dt}$, $\dfrac{d\theta}{dt}$.

By means of these values of the first differentials, and neglecting $\dfrac{d^3r}{dt^3}$ and $\dfrac{d^4r}{dt^4}$ in (4), we find $\dfrac{d^2r}{dt^2}$ from (4).

Returning to (2) and (3) and using this value of $\dfrac{d^2r}{dt^2}$, we re-determine the first differentials, and repeat the process until the final values of $\dfrac{dr}{dt}$ and $\dfrac{d\theta}{dt}$ remain unchanged. We thus obtain the velocity at this point r, θ on the supposition that $\dfrac{d^3r}{dt^3}$, $\dfrac{d^4r}{dt^4}$ are negligible, and on substitution in the last of (1) we obtain the value of C corresponding to the orbit which passes through the chosen point.

Brown then shews how the remainder of the orbit may be traced with all desirable accuracy in the case where ν is large. It does not concern me to follow him here, since his process could scarcely be applicable for small values of ν. But if his scheme should still lead to the required result, the remainder of the orbit might be traced by quadratures, and this is the plan which I have adopted. If the orbit as so determined proves to be clearly non-periodic, it seems safe to conclude that no widely librating planets can exist for small values of ν.

I had already become fairly confident from a number of trials, which will be referred to hereafter, that such orbits do not exist; but it seemed worth while to make one more attempt by Brown's procedure, and the result appears to be of sufficient interest to be worthy of record.

For certain reasons I chose as my starting-point

$$x_0 = -\cdot36200, \quad y_0 = \cdot93441,$$

which give $\qquad r_0 = 1\cdot00205, \quad \rho_0 = 1\cdot65173.$

The successive approximations to C were found to be

$$33{\cdot}6977, \quad 33{\cdot}7285, \quad 33{\cdot}7237, \quad 33{\cdot}7246, \quad 33{\cdot}7243.$$

I therefore took the last value as that of C, and found also that the direction of motion was given by $\phi_0 = 2^\circ\ 21'$. These values of x_0, y_0, ϕ_0, and C then furnish the values from which to begin the quadratures.

Fig. 1 shews the result, the starting-point being at B. The curve was traced backwards to A and onwards to C, and the computed positions are shewn by dots connected into a sweeping curve by dashes.

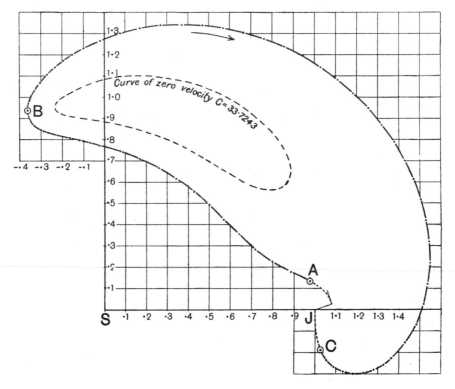

Fig. 1. Results derived from Professor Brown's Method.

From A back to perijove and from C on to J the orbit was computed as undisturbed by the Sun*. Within the limits of accuracy adopted the body collides with J.

* When the body has been traced to the neighbourhood of J, let it be required to determine its future position on the supposition that the solar perturbation is negligible. Since the axes of reference are rotating, the solution needs care, and it may save the reader some trouble if I set down how it may be done conveniently.

Let the coordinates, direction of motion, and velocity, at the moment $t=0$ when solar perturbation is to be neglected, be given by x_0, y_0 (or r_0, θ_0, and ρ_0, ψ_0), ϕ_0, V_0; and generally

Since the curve comes down on to the negative side of the line of syzygy SJ it differs much from Brown's orbits, and it is clear that it is not periodic. Thus his method fails, and there is good reason to believe that his conjecture is unfounded.

After this work had been done Professor Brown pointed out to me in a letter that if his process be translated into rectangular coordinates, the approximate expressions for dx/dt and dy/dt will have as a divisor the function

$$Q = \left(4n^2 - \frac{\partial^2 \Omega}{\partial x^2}\right)\left(4n^2 - \frac{\partial^2 \Omega}{\partial y^2}\right) - \left(\frac{\partial^2 \Omega}{\partial x \partial y}\right)^2.$$

The method will then fail if Q vanishes or is small.

let the suffix 0 to any symbol denote its value at this epoch. Then the mean distance a, mean motion μ, and eccentricity e are found from

$$\frac{1}{a} = \frac{2}{\rho_0} - [V_0^2 + 2n\rho_0 V_0 \cos (\phi_0 - \psi_0) + n^2\rho_0^2], \quad \mu^2 a^3 = 1,$$
$$a (1 - e^2) = [V_0 \rho_0 \cos (\phi_0 - \psi_0) + n\rho_0^2]^2.$$

Let $t = \tau$ be the time of passage of perijove, so that when τ is positive perijove is later than the epoch $t = 0$.

At any time t let ρ, v, E be radius vector, true and eccentric anomalies; then

$$\rho = a (1 - e \cos E),$$
$$\rho^{\frac{1}{2}} \cos \tfrac{1}{2}v = a^{\frac{1}{2}} (1 - e)^{\frac{1}{2}} \cos \tfrac{1}{2}E,$$
$$\rho^{\frac{1}{2}} \sin \tfrac{1}{2}v = a^{\frac{1}{2}} (1 + e)^{\frac{1}{2}} \sin \tfrac{1}{2}E,$$
$$\mu (t - \tau) = E - e \sin E,$$
$$\psi = \psi_0 - v_0 + v - nt.$$

On putting $t = 0$, E_0 and τ may be computed from these formulae, and it must be noted that when τ is positive E_0 and v_0 are to be taken as negative.

The position of the body as it passes perijove is clearly given by

$$x - 1 = a (1 - e) \cos (\psi_0 - v_0 - n\tau), \quad y = a (1 - e) \sin (\psi_0 - v_0 - n\tau).$$

Any other position is to be found by assuming a value for E, computing ρ, v, t, ψ, and using the formulae

$$x - 1 = \rho \cos \psi, \quad y = \rho \sin \psi.$$

In order to find V and ϕ we require the formulae

$$\frac{1}{\rho}\frac{d\rho}{dt} = \frac{ae \sin E}{\rho} \cdot \frac{\mu a}{\rho}; \quad \frac{dv}{dt} = \frac{[a (1 - e^2)]^{\frac{1}{2}}}{\rho} \cdot \frac{a^{\frac{1}{2}}}{\rho} \cdot \frac{\mu a}{\rho},$$

and

$$V \sin \phi = -\frac{(x - 1)}{\rho}\frac{d\rho}{dt} + y \left(\frac{dv}{dt} - n\right),$$

$$V \cos \phi = \frac{y}{\rho}\frac{d\rho}{dt} + (x - 1) \left(\frac{dv}{dt} - n\right).$$

The value of V as computed from these should be compared with that derived from

$$V^2 = \nu \left(r^2 + \frac{2}{r}\right) + \left(\rho^2 + \frac{2}{\rho}\right) - C,$$

and if the two agree pretty closely, the assumption as to the insignificance of solar perturbation is justified.

If the orbit is retrograde about J, care has to be taken to use the signs correctly, for v and E will be measured in a retrograde direction, whereas ψ will be measured in the positive direction.

A similar investigation is applicable, *mutatis mutandis*, when the body passes very close to S.

I find that if we write $\Gamma = \dfrac{\nu}{r^3} + \dfrac{1}{\rho^3}$, the divisor may be written in the form

$$Q = (3n^2 + \Gamma)(3\dot{n^2} - 2\Gamma) + \frac{\rho\nu}{r^5\rho^5} \sin\theta \sin\psi.$$

Now, Mr T. H. Brown, Professor Brown's pupil, has traced one portion of the curve $Q = 0$, corresponding to $\nu = 10$, and he finds that it passes rather near to the orbit I have traced. This confirms the failure of the method which I had concluded otherwise.

§ 2. *Variation of an Orbit.*

A great difficulty in determining the orbits of librating planets by quadratures arises from the fact that these orbits do not cut the line of syzygies at right angles, and therefore the direction of motion is quite indeterminate at every point. I endeavoured to meet this difficulty by a method of variation which is certainly feasible, but, unfortunately, very laborious. In my earlier attempts I had drawn certain orbits, and I attempted to utilise the work by the method which will now be described.

The stability of a periodic orbit is determined by varying the orbit. The form of the differential equation which the variation must satisfy does not depend on the fact that the orbit is periodic, and thus the investigation in §§ 8, 9 of my paper on " Periodic Orbits " remains equally true when the varied orbit is not periodic.

Suppose, then, that the body is displaced from a given point of a non-periodic orbit through small distances $\delta q\, V^{-\frac{1}{2}}$ along the outward normal and δs along the positive tangent, then we must have

$$\frac{d^2\delta q}{ds^2} + \Psi\delta q = 0,$$

$$\frac{d}{ds}\left(\frac{\delta s}{V}\right) = -\frac{2\delta q}{V^{\frac{3}{2}}}\left(\frac{1}{R} + \frac{n}{V}\right),$$

where

$$\Psi = \frac{5}{2}\left(\frac{1}{R} + \frac{n}{V}\right)^2 - \frac{3}{2V^2}\left[\frac{\nu}{r^3}\cos^2(\phi - \theta) + \frac{1}{\rho^3}\cos^2(\phi - \psi)\right] + \frac{3}{4}\left(\frac{dV}{Vds}\right)^2,$$

and

$$\frac{dV}{Vds} = \frac{\nu}{V^2}\left(\frac{1}{r^2} - r\right)\sin(\phi - \theta) + \frac{1}{V^2}\left(\frac{1}{\rho^2} - \rho\right)\sin(\phi - \psi).$$

Also

$$\delta\phi = -\frac{1}{V^{\frac{1}{2}}}\left[\frac{d\delta q}{ds} - \tfrac{1}{2}\delta q\left(\frac{dV}{Vds}\right)\right] + \frac{\delta s}{R}.$$

Since it is supposed that the coordinates, direction of motion, and radius of curvature R have been found at a number of points equally distributed along the orbit, it is clear that Ψ may be computed for each of those points.

At the point chosen as the starting-point the variation may be of two kinds:—

(1) $\delta q_0 = a$, $\dfrac{d\delta q_0}{ds} = 0$, where a is a constant,

(2) $\delta q_0 = 0$, $\dfrac{d\delta q_0}{ds} = b$, where b is a constant.

Each of these will give rise to an independent solution, and if in either of them a or b is multiplied by any factor, that factor will multiply all the succeeding results. It follows, therefore, that we need not concern ourselves with the exact numerical values of a or b, but the two solutions will give us all the variations possible. In the first solution we start parallel with the original curve at the chosen point on either side of it, and at any arbitrarily chosen small distance. In the second we start from the chosen point, but at any arbitrary small inclination on either side of the original tangent.

The solution of the equations for δq and δs have to be carried out step by step along the curve, and it may be worth while to indicate how the work may be arranged.

The length of arc from point to point of the unvaried orbit may be denoted by Δs, and we may take four successive values of Ψ, say Ψ_{n-1}, Ψ_n, Ψ_{n+1}, Ψ_{n+2}, as affording a sufficient representation of the march of the function Ψ throughout the arc Δs between the points indicated by n to $n+1$.

If the differential equation for δq be multiplied by $(\Delta s)^2$, and if we introduce a new independent variable z such that $dz = ds/\Delta s$, and write $X = \Psi (\Delta s)^2$, the equation becomes

$$\frac{d^2 \delta q}{dz^2} = - X \delta q,$$

and z increases by unity as the arc increases by Δs.

Suppose that the integration has been carried as far as the point n, and that δq_0, $d\delta q_0/dz$ are the values at that point; then it is required to find δq_1, $d\delta q_1/dz$ at the point $n+1$.

If the four adjacent values of X are X_{-1}, X_0, X_1, X_2, and if

$$\delta_1 = X_1 - X_0, \quad \delta_2 = \tfrac{1}{2}\left[(X_2 - 2X_1 + X_0) + (X_1 - 2X_0 + X_{-1})\right],$$

Bessel's formula for the function X is

$$X = X_0 + (\delta_1 - \tfrac{1}{2}\delta_2) z + \tfrac{1}{2}\delta_2 z^2$$

We now assume that throughout the arc n to $n+1$,

$$\delta q = \delta q_0 + \frac{d\delta q_0}{dz}z + Q_2 z^2 + Q_3 z^3 + Q_4 z^4,$$

where Q_2, Q_3, Q_4 have to be determined so as to satisfy the differential equation.

On forming the product $X\delta q$, integrating, and equating coefficients, we find $Q_2 = -\frac{1}{2}X_0\delta q_0$, and the values of Q_3, Q_4 are easily found. In carrying out this work I neglect all terms of the second order except X_0^2.

The result may be arranged as follows:—

Let
$$A = 1 - \tfrac{1}{2}X_0 - \tfrac{1}{6}\delta_1 + \tfrac{1}{24}(\delta_2 + X_0^2),$$
$$B = 1 - \tfrac{1}{6}X_0 - \tfrac{1}{12}\delta_1 + \tfrac{1}{24}\delta_2,$$
$$C = X_0 + \tfrac{1}{2}\delta_1 + \tfrac{1}{12}\delta_2 - \tfrac{1}{6}(\delta_2 + X_0^2),$$
$$D = 1 - \tfrac{1}{2}X_0 - \tfrac{1}{3}\delta_1 + \tfrac{1}{6}\delta_2;$$

then, on putting $z = 1$, we find

$$\delta q_1 = A\delta q_0 + B\frac{d\delta q_0}{dz},$$

$$\frac{d\delta q_1}{dz} = -C\delta q_0 + D\frac{d\delta q_0}{dz}.$$

When the Ψ's have been computed, the X's and A, B, C, D are easily found at each point of the unvaried orbit. We then begin the two solutions from the chosen starting-point, and thus trace δq and $d\delta q/dz$ from point to point both backwards and forwards. The necessary change of procedure when Δs changes in magnitude is obvious.

The procedure is tedious although easy, but the work is enormously increased when we pass on further to obtain an intelligible result from the integration. When δq and $d\delta q/dz$ have been found at each point, a further integration has to be made to determine δs, and this has, of course, to be done for each of the solutions. Next, we have to find the normal displacement δp (equal to $\delta q V^{-\frac{1}{2}}$), and, finally, δp, δs have to be converted into rectangular displacements δx, δy.

The whole process is certainly very laborious; but when the result is attained it does furnish a great deal of information as to the character of the orbits adjacent to the orbit chosen for variation. I only carried the work through in one case, because I had gained enough information by this single instance. However, it does not seem worth while to record the numerical results in that case.

In the variation which has been described, C is maintained unchanged, but it is also possible to vary C. If C becomes $C + \delta C$, it will be found that the equations assume the form

$$\frac{d^2 \delta q}{ds^2} + \Psi \delta q + \frac{\delta C}{V^{\frac{3}{2}}} \left(\frac{1}{R} + \frac{n}{V} \right) = 0,$$

$$\frac{d}{ds} \left(\frac{\delta s}{V} \right) + \frac{2 \delta q}{V^{\frac{3}{2}}} \left(\frac{1}{R} + \frac{n}{V} \right) + \frac{\delta C}{2V^2} = 0.$$

But this kind of variation cannot be used with much advantage, for although it is possible to evaluate δq and δs for specific initial values of δC, δq, $d\delta q/ds$ at a specific initial point, only one single varied orbit is so deducible. In the previous case we may assign any arbitrary values, either positive or negative, to the constants denoted by a and b, and thus find a group of varied orbits.

§ 3. *A New Family of Periodic Orbits.*

In attempting to discover an example of an orbit of the kind suspected by Brown, I traced a number of orbits. Amongst these was that one which was varied as explained in § 2, although when the variation was effected I did not suspect it to be in reality periodic in a new way. It was clear that it could not be one of Brown's orbits, and I therefore put the work aside and made a fresh attempt, as explained in § 1. Finally, for my own satisfaction, I completed the circuit of this discarded orbit, and found to my surprise that it belonged to a new and unsuspected class of periodic. The orbit in question is that marked 33·5 in fig. 3, where only the half of it is drawn which lies on the positive side of SJ.

It will be convenient to use certain terms to indicate the various parts of the orbits under discussion, and these will now be explained. Periodic orbits have in reality neither beginning nor end; but, as it will be convenient to follow them in the direction traversed from an orthogonal crossing of the line of syzygies, I shall describe the first crossing as the "beginning" and the second orthogonal crossing of SJ as the "end." I shall call the large curve surrounding the apex of the Lagrangian equilateral triangle the "loop," and this is always described in the clockwise or negative direction. The portions of the orbit near J will be called the "circuit," or the "half-" or "quarter-circuit," as the case may be. The "half-circuits" about J are described counter-clockwise or positively, but where there is a complete "circuit" it is clockwise or negative. For example, in fig. 3 the orbit 33·5 "begins" with a positive quarter-circuit, passes on to a negative "loop," and "ends" in a positive quarter-circuit. Since the initial and final quarter-circuits both cut SJ at right angles, the orbit is periodic, and would be completed by a similar curve on the negative side of SJ. In the completed orbit positively described half-circuits are interposed between negative loops described alternately on the positive and negative sides of SJ.

Having found this orbit almost by accident, it was desirable to find other orbits of this kind; but the work was too heavy to obtain as many as is desirable. There seems at present no way of proceeding except by conjecture, and bad luck attended the attempts to draw the curve when C is 33·25. The various curves are shewn in fig. 2, from which this orbit may be constructed with substantial accuracy.

In fig. 2 the firm line of the external loop was computed backwards, starting at right angles to SJ from $x = ·95$, $y = 0$, the point to which 480° is attached. After the completion of the loop, the curve failed to come down

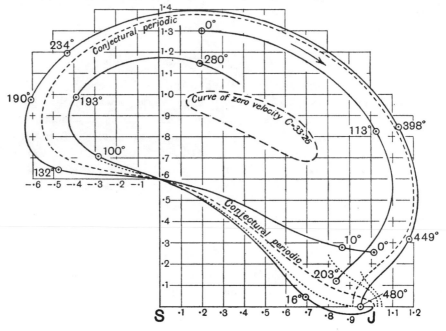

Fig. 2.   Orbits computed for the Case of $C = 33·25$.

close to J as was hoped, but came to the points marked 10° and 0°. The "beginnings" of two positively described quarter-circuits about J are shewn as dotted lines, and an orbit of ejection, also dotted, is carried somewhat further. Then there is an orbit, shewn in firm line, "beginning" with a negative half-circuit about J, and when this orbit had been traced half-way through its loop it appeared that the body was drawing too near to the curve of zero velocity, from which it would rebound, as one may say. This orbit is continued in a sense by a detached portion starting from a horizontal tangent at $x = ·2$, $y = 1·3$. It became clear ultimately that the horizontal tangent ought to have been chosen with a somewhat larger value for y. From these attempts it may be concluded that the periodic orbit must resemble the broken line marked as conjectural, and as such it is transferred to fig. 3 and shewn there as a dotted curve.   I shall return hereafter to the explanation of the degrees written along these curves.

Much better fortune attended the construction of the orbit 33·75 shewn in fig. 3, for, although the final perijove does not fall quite on the line of syzygies, yet the true periodic orbit can differ but little from that shewn. It will be noticed that in this case the orbit "ends" with a negative half-circuit, and it is thus clear that if we were to watch the march of these

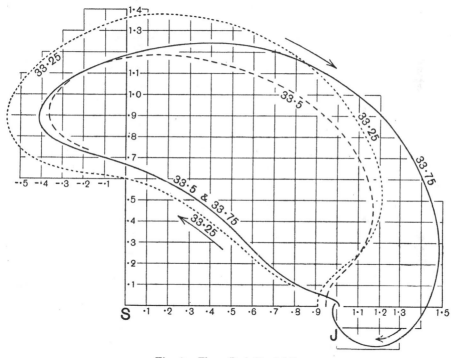

Fig. 3.   Three Periodic Orbits.

orbits as C falls from 33·75 to 33·5 we should see the negative half-circuit shrink, pass through the ejectional stage, and emerge as a positive quarter-circuit when C is 33·5.

The three orbits shewn in fig. 3 are the only members of this family that I have traced.   It will be noticed that they do not exhibit that regular progress from member to member which might have been expected from the fact that the values of C are equidistant from one another.   It might be suspected that they are really members of different families presenting similar characteristics, but I do not think this furnishes the explanation.

In describing the loop throughout most of its course the body moves roughly parallel to the curve of zero velocity. For the values of C involved here that curve is half of the broken horse-shoe described in my paper on "Periodic Orbits" (*Scientific Papers*, vol. IV., p. 11, or *Acta Math.*, vol. XXI. (1897)). Now, for $\nu = 10$ the horse-shoe breaks when C has fallen to 34·91, and below that value each half of the broken horse-shoe, which delimits the forbidden space, shrinks. Now, since the orbits follow the contour of the horse-shoe, it might be supposed that the orbits would also shrink as C falls in magnitude. On the other hand, as C falls from 33·5 to 33·25, our figures shew that the loop undoubtedly increases in size. This latter consideration would lead us to conjecture that the loop for 33·75 should be smaller than that for 33·5. Thus, looking at the matter from one point of view, we should expect the orbits to shrink, and from another to swell as C falls in value. It thus becomes intelligible that neither conjecture can be wholly correct, and we may thus find an explanation of the interlacing of the orbits as shewn in my fig. 3.

It is certain from general considerations that families of orbits must originate in pairs, and we must therefore examine the origin of these orbits, and consider the fate of the other member of the pair.

It may be that for values of C greater than 33·75 the initial positive quarter-circuit about J is replaced by a negative half-circuit; but it is unnecessary for the present discussion to determine whether this is so or not, and it will suffice to assume that when C is greater than 33·75 the "beginning" is as shewn in my figure. The "end" of 33·75 is a clearly marked negative half-circuit, and this shews that the family originates from a coalescent pair of orbits "ending" in such a negative half-circuit, with identical final orthogonal crossing of SJ in which the body passes from the negative to the positive side of SJ.

This coalescence must occur for some critical value of C between 34·91 and 33·75, and it is clear that as C falls below that critical value one of the "final" orthogonal intersections will move towards S and the other towards J.

In that one of the pair for which the intersection moves towards S the negative circuit increases in size; in the other in which it moves towards J the circuit diminishes in size, and these are clearly the orbits which have been traced. We next see that the negative circuit vanishes, the orbit becomes ejectional, and the motion about J both at "beginning" and "end" has become positive.

It may be suspected that when C falls below 33·25 the half-circuits round J increase in magnitude, and that the orbit tends to assume the form of a sort of asymmetrical double figure-of-8, something like the figure which Lord Kelvin drew as an illustration of his graphical method of curve-tracing*.

In the neighbourhood of Jove the motion of the body is rapid, but the loops are described very slowly. The number of degrees written along the curves in fig. 2 represent the angles turned through by Jove about the Sun since the moment corresponding to the position marked 0°. Thus the firm line which lies externally throughout most of the loop terminates with 480°. Since this orbit cuts SJ orthogonally, it may be continued symmetrically on the negative side of SJ, and therefore while the body moves from the point 0° to a symmetrical one on the negative side Jove has turned through 960° round the Sun, that is to say, through $2\frac{2}{3}$ revolutions.

Again, in the case of the orbit beginning with a negative half-circuit, shewn as a firm line, Jove has revolved through 280° by the time the point so marked is reached. We may regard this as continued in a sense by the detached portion of an orbit marked with 0°, 113°, 203°; and since 280° + 203° is equal to 483°, we again see that the period of the periodic orbit must be about 960°, or perhaps a little more.

In the cases of the other orbits more precise values may be assigned. For $C = 33\cdot5$, the angle nT (where T is the period) is 1115° or 3·1 revolutions of Jove; and for $C = 33\cdot75$, nT is 1235° or 3·4 revolutions.

It did not seem practicable to investigate the stability of these orbits, but we may suspect them to be unstable.

The numerical values for drawing the orbits $C = 33\cdot5$ and $33\cdot75$ are given in an appendix, but those for the various orbits from which the conjectural orbit $C = 33\cdot25$ is constructed are omitted. I estimate that it is as laborious to trace one of these orbits as to determine fully half a dozen of the simpler orbits shewn in my earlier paper.

Although the present contribution to our knowledge is very imperfect, yet it may be hoped that it will furnish the mathematician with an intimation worth having as to the orbits towards which his researches must lead him.

The librating planets were first recognised as small oscillations about the triangular positions of Lagrange, and they have now received a very remarkable extension at the hands of Professor Brown. It appears to me that the family of orbits here investigated possesses an interesting relationship to these librating planets, for there must be orbits describing double, triple, and multiple loops in the intervals between successive half-circuits about Jove. Now, a body which describes its loop an infinite number of times,

* *Popular Lectures*, vol. i., 2nd ed., p. 31; *Phil. Mag.*, vol. xxxiv., 1892, p. 443.

before it ceases to circulate round the triangular point, is in fact a librating planet. It may be conjectured that when the Sun's mass ν is yet smaller than 10, no such orbit as those traced is possible. When ν has increased to 10, probably only a single loop is possible; for a larger value a double loop may be described, and then successively more frequently described multiple loops will be reached. When ν has reached 24·9599 a loop described an infinite number of times must have become possible, since this is the smallest value of ν which permits oscillation about the triangular point. If this idea is correct, and if N denotes the number expressing the multiplicity of the loop, then as ν increases $dN/d\nu$ must tend to infinity; and I do not see why this should not be the case.

These orbits throw some light on cosmogony, for we see how small planets with the same mean motion as Jove in the course of their vicissitudes tend to pass close to Jove, ultimately to be absorbed into its mass. We thus see something of the machinery whereby a large planet generates for itself a clear space in which to circulate about the Sun.

My attention was first drawn to periodic orbits by the desire to discover how a Laplacian ring could coalesce into a planet. With that object in view I tried to discover how a large planet would affect the motion of a small one moving in a circular orbit at the same mean distance. After various failures the investigation drifted towards the work of Hill and Poincaré, so that the original point of view was quite lost and it is not even mentioned in my paper on "Periodic Orbits." It is of interest, to me at least, to find that the original aspect of the problem has emerged again.

APPENDIX.

Numerical results of Quadratures.

$$C = 33\cdot5.$$

Perijove $x_0 = 1\cdot0171$, $y_0 = -\cdot0034$, taken as zero.
Time from perijove up to $s = -2\cdot1$ is given by $nt = 9°\ 25'$.

s	x	y	ϕ	$2n/V$
$-2\cdot1$	$+\ \cdot8282$	$+\ \cdot0980$	$+66°\ 10'$	$2\cdot408$
$2\cdot0$	$\cdot7409$	$\cdot1467$	$55\ \ 53$	$2\cdot829$
$1\cdot9$	$\cdot6625$	$\cdot2084$	$48\ \ 36$	$2\cdot876$
$1\cdot8$	$\cdot5894$	$\cdot2766$	$46\ \ \ 3$	$2\cdot768$
$1\cdot7$	$\cdot5171$	$\cdot3457$	$46\ \ 55$	$2\cdot655$
$1\cdot6$	$\cdot4425$	$\cdot4124$	$49\ \ 46$	$2\cdot584$
$1\cdot5$	$\cdot3641$	$\cdot4744$	$53\ \ 39$	$2\cdot568$
$-1\cdot4$	$+\ \cdot2814$	$+\ \cdot5306$	$+57\ \ 56$	$2\cdot613$

s	x	y	ϕ	$2n/V$
			° ′	
$-1\cdot3$	$+\ \cdot1948$	$+\ \cdot5805$	$+62\quad 8$	$2\cdot728$
$1\cdot2$	$\cdot1049$	$\cdot6243$	$65\ 51$	$2\cdot930$
$1\cdot1$	$+\ \cdot0126$	$\cdot6628$	$68\ 38$	$3\cdot251$
$1\cdot0$	$-\ \cdot0810$	$\cdot6979$	$69\ 46$	$3\cdot760$
$\cdot9$	$\cdot1747$	$\cdot7330$	$68\quad 7$	$4\cdot598$
$\cdot85$	$\cdot2207$	$\cdot7526$	$65\ 13$	$5\cdot240$
$\cdot8$	$\cdot2653$	$\cdot7754$	$60\quad 1$	$6\cdot133$
$\cdot75$	$\cdot3068$	$\cdot8035$	$50\ 51$	$7\cdot377$
$\cdot725$	$\cdot3252$	$\cdot8203$	$44\quad 2$	$8\cdot139$
$\cdot7$	$\cdot3412$	$\cdot8395$	$35\ 17$	$8\cdot944$
$\cdot675$	$\cdot3537$	$\cdot8611$	$24\ 33$	$9\cdot664$
$\cdot65$	$\cdot3617$	$\cdot8848$	$12\ 27$	$10\cdot129$
$\cdot625$	$\cdot3644$	$\cdot9096$	$+\ 0\ 13$	$10\cdot224$
$\cdot6$	$\cdot3620$	$\cdot9344$	$-10\ 56$	$10\cdot009$
$\cdot575$	$\cdot3552$	$\cdot9584$	$20\ 31$	$9\cdot655$
$\cdot55$	$\cdot3448$	$\cdot9811$	$28\ 30$	$9\cdot205$
$\cdot5$	$\cdot3161$	$1\cdot0220$	$40\ 48$	$8\cdot448$
$\cdot45$	$\cdot2806$	$1\cdot0571$	$49\ 38$	$7\cdot872$
$\cdot4$	$\cdot2405$	$1\cdot0869$	$56\ 51$	$7\cdot460$
$\cdot3$	$\cdot1518$	$1\cdot1326$	$68\quad 4$	$6\cdot961$
$\cdot2$	$-\ \cdot0565$	$1\cdot1626$	$76\ 47$	$6\cdot730$
$-\ \cdot1$	$+\ \cdot0421$	$1\cdot1791$	$83\ 58$	$6\cdot647$
$\cdot0$	$\cdot1419$	$1\cdot1842$	$-90\quad 0$	$6\cdot633$
$+\ \cdot05$	$\cdot1919$	$1\cdot1830$	$180°+87\ 21$	$6\cdot630$
$\cdot1$	$\cdot2418$	$1\cdot1797$	$84\ 54$	$6\cdot626$
$\cdot15$	$\cdot2915$	$1\cdot1742$	$82\ 38$	$6\cdot609$
$\cdot2$	$\cdot3410$	$1\cdot1669$	$80\ 31$	$6\cdot572$
$\cdot3$	$\cdot4389$	$1\cdot1470$	$76\ 31$	$6\cdot432$
$\cdot4$	$\cdot5353$	$1\cdot1203$	$72\ 33$	$6\cdot201$
$\cdot5$	$\cdot6295$	$1\cdot0869$	$68\ 16$	$5\cdot912$
$\cdot6$	$\cdot7208$	$1\cdot0461$	$63\ 29$	$5\cdot605$
$\cdot7$	$\cdot8081$	$\cdot9974$	$58\quad 8$	$5\cdot313$
$\cdot8$	$\cdot8902$	$\cdot9404$	$52\ 12$	$5\cdot055$
$\cdot9$	$\cdot9656$	$\cdot8748$	$45\ 39$	$4\cdot842$
$1\cdot0$	$1\cdot0326$	$\cdot8006$	$38\ 22$	$4\cdot671$
$1\cdot1$	$1\cdot0889$	$\cdot7181$	$30\ 11$	$4\cdot540$
$1\cdot2$	$1\cdot1321$	$\cdot6280$	$20\ 46$	$4\cdot435$
$1\cdot3$	$1\cdot1585$	$\cdot5318$	$9\ 38$	$4\cdot326$
$1\cdot35$	$1\cdot1642$	$\cdot4821$	$180°+\ 3\ 16$	$4\cdot250$
$1\cdot4$	$1\cdot1641$	$\cdot4322$	$180°-\ 3\ 40$	$4\cdot141$
$1\cdot45$	$1\cdot1577$	$\cdot3826$	$11\quad 5$	$3\cdot983$
$1\cdot5$	$1\cdot1448$	$\cdot3343$	$18\ 44$	$3\cdot758$
$1\cdot55$	$1\cdot1257$	$\cdot2881$	$26\quad 8$	$3\cdot460$
$1\cdot6$	$1\cdot1011$	$\cdot2446$	$32\ 39$	$3\cdot100$
$1\cdot65$	$1\cdot0723$	$\cdot2038$	$37\ 33$	$2\cdot701$
$1\cdot7$	$1\cdot0408$	$\cdot1650$	$40\quad 4$	$2\cdot291$
$+1\cdot75$	$+1\cdot0087$	$+\ \cdot1267$	$180°-39\ 12$	$1\cdot893$

Time from $s=1\cdot75$ to perijove given by $nt=5°\ 58'$.

Coordinates of perijove $x=\cdot9501$, $y=-\cdot0029$.

The following additional positions were calculated backwards from a perijove at $x = \cdot95$, $y=0$, $\phi=180°$.

x	y	ϕ
$+ \cdot9500$	$+ \cdot0000$	$180° +$ 0° 0′
$\cdot9512$	$\cdot0531$	$180° -$ 22 30
$\cdot9647$	$\cdot0797$	30 52
$\cdot9756$	$\cdot0966$	34 48
$\cdot9874$	$\cdot1127$	37 37
$1\cdot0128$	$\cdot1436$	40 37
$1\cdot0390$	$\cdot1738$	40 56
$1\cdot0649$	$\cdot2043$	39 12
$1\cdot0893$	$\cdot2360$	35 51
$1\cdot1114$	$\cdot2693$	31 16
$1\cdot1463$	$\cdot3412$	20 10
$+1\cdot1661$	$+ \cdot4186$	$180° -$ 8 40

This supplementary orbit becomes indistinguishable in a figure of moderate size from the preceding orbit, which is therefore accepted as being periodic. The period is given by $nT = 1115° \cdot 4 = 3 \cdot 1$ revolutions of Jove.

$$C = 33 \cdot 75.$$

This orbit was computed from a conjectural starting-point which seemed likely to lead to the desired result; the computation was finally carried backwards from the starting-point. The coordinates of perijove were found to be $x_0 = 1 \cdot 0106$, $y_0 = \cdot 0006$, which may be taken as virtually on the line of syzygies. The motion from perijove is direct.

s	x	y	ϕ	$2n/V$
perijove	$+1\cdot0106$	$+ \cdot0006$	0° 0′ very nearly	
$- \cdot35$	$\cdot9652$	$\cdot0403$	66 38	$1\cdot140$
$- \cdot3$	$\cdot9184$	$\cdot0578$	71 6	$1\cdot635$
$- \cdot25$	$\cdot8713$	$\cdot0744$	69 27	$2\cdot075$
$- \cdot2$	$\cdot8251$	$\cdot0936$	65 3	$2\cdot447$
$- \cdot1$	$\cdot7391$	$\cdot1444$	54 15	$2\cdot882$
$0\cdot0$	$\cdot6625$	$\cdot2084$	47 0	$2\cdot946$
$\cdot1$	$\cdot5911$	$\cdot2785$	44 44	$2\cdot850$
$\cdot2$	$\cdot5202$	$\cdot3490$	46 0	$2\cdot749$
$\cdot3$	$\cdot4465$	$\cdot4165$	49 13	$2\cdot686$
$\cdot4$	$\cdot3685$	$\cdot4791$	53 29	$2\cdot675$
$\cdot5$	$\cdot2858$	$\cdot5352$	58 10	$2\cdot723$
$\cdot6$	$\cdot1987$	$\cdot5844$	62 52	$2\cdot838$
$\cdot7$	$\cdot1081$	$\cdot6265$	67 13	$3\cdot036$
$\cdot8$	$+ \cdot0147$	$\cdot6622$	70 49	$3\cdot348$
$\cdot9$	$- \cdot0805$	$\cdot6929$	73 11	$3\cdot834$
$1\cdot0$	$\cdot1764$	$\cdot7213$	73 25	$4\cdot631$
$1\cdot1$	$\cdot2713$	$\cdot7525$	69 17	$6\cdot090$
$1\cdot15$	$\cdot3173$	$\cdot7721$	63 50	$7\cdot333$
$1\cdot2$	$\cdot3601$	$\cdot7977$	53 25	$9\cdot236$
$1\cdot225$	$\cdot3791$	$\cdot8140$	45 6	$10\cdot360$
$1\cdot25$	$- \cdot3951$	$+ \cdot8332$	33 54	$11\cdot840$

s	x	y	ϕ	$2n/V$
			° '	
1·275	− ·4064	+ ·8553	+19 53	12·955
1·3	·4118	·8796	+ 4 42	13·412
1·325	·4108	·9046	− 9 14	13·174
1·35	·4043	·9287	20 35	12·599
1·375	·3936	·9513	29 25	11·945
1·4	·3800	·9723	36 21	11·364
1·45	·3466	1·0096	46 23	10·471
1·5	·3082	1·0416	53 25	9·849
1·6	·2227	1·0940	62 21	9·034
1·7	·1317	1·1356	67 59	8·347
1·8	− ·0377	1·1696	72 2	7·618
2·0	+ ·1563	1·2184	79 17	6·140
2·2	·3547	1·2407	−88 13	4·966
2·4	·5541	1·2300	180°+81 54	4·182
2·6	·7487	1·1845	71 49	3·665
2·8	·9322	1·1057	61 40	3·305
3·0	1·0989	·9956	51 24	3·052
3·2	1·2429	·8573	40 54	2·873
3·4	1·3588	·6946	29 55	2·751
3·6	1·4402	·5123	18 1	2·682
3·8	1·4797	·3168	180°+ 4 28	2·670
4·0	1·4674	·1181	180°− 12 14	2·733
4·1	1·4377	+ ·0227	23 43	2·806
4·2	1·3894	− ·0646	35 38	2·910
4·3	1·3208	·1366	52 23	3·027
4·35	1·2787	·1635	62 47	3·068
4·4	1·2322	·1817	74 47	3·063
4·45	1·1829	·1892	180°− 88 15	2·983
4·5	1·1332	·1845	+77 25	2·780
4·55	1·0863	·1676	63 8	2·477
4·6	1·0448	·1399	49 32	2·101
4·65	1·0108	·1034	36 18	1·683
4·7	·9867	− ·0598	21 1	1·234
perijove	+ ·990	+ ·011	about 49	

The orbit is not rigorously periodic, but an extremely small change at the beginning would make it so. The period is given by $nT = 1234°·6 = 3·43$ revolutions of Jove.

ADDRESS

(DELIVERED BEFORE THE INTERNATIONAL CONGRESS OF MATHEMATICIANS
AT CAMBRIDGE IN 1912)

FOUR years ago at our Conference at Rome the Cambridge Philosophical
Society did itself the honour of inviting the International Congress of
Mathematicians to hold its next meeting at Cambridge. And now I, as
President of the Society, have the pleasure of making you welcome here.
I shall leave it to the Vice-Chancellor, who will speak after me, to express
the feeling of the University as a whole on this occasion, and I shall
confine myself to my proper duty as the representative of our Scientific
Society.

The Science of Mathematics is now so wide and is already so much
specialised that it may be doubted whether there exists to-day any man
fully competent to understand mathematical research in all its many diverse
branches. I, at least, feel how profoundly ill-equipped I am to represent
our Society as regards all that vast field of knowledge which we classify as
pure mathematics. I must tell you frankly that when I gaze on some of the
papers written by men in this room I feel myself much in the same position
as if they were written in Sanskrit.

But if there is any place in the world in which so one-sided a President
of the body which has the honour to bid you welcome is not wholly out of
place it is perhaps Cambridge. It is true that there have been in the past
at Cambridge great pure mathematicians such as Cayley and Sylvester, but
we surely may claim without undue boasting that our University has played
a conspicuous part in the advance of applied mathematics. Newton was
a glory to all mankind, yet we Cambridge men are proud that fate ordained
that he should have been Lucasian Professor here. But as regards the part
played by Cambridge I refer rather to the men of the last hundred years,
such as Airy, Adams, Maxwell, Stokes, Kelvin, and other lesser lights, who
have marked out the lines of research in applied mathematics as studied in
this University. Then too there are others such as our Chancellor, Lord
Rayleigh, who are happily still with us.

Up to a few weeks ago there was one man who alone of all mathematicians might have occupied the place which I hold without misgivings as to his fitness; I mean Henri Poincaré. It was at Rome just four years ago that the first dark shadow fell on us of that illness which has now terminated so fatally. You all remember the dismay which fell on us when the word passed from man to man "Poincaré is ill." We had hoped that we might again have heard from his mouth some such luminous address as that which he gave at Rome; but it was not to be, and the loss of France in his death affects the whole world.

It was in 1900 that, as president of the Royal Astronomical Society, I had the privilege of handing to Poincaré the medal of the Society, and I then attempted to give an appreciation of his work on the theory of the tides, on figures of equilibrium of rotating fluid and on the problem of the three bodies. Again in the preface to the third volume of my collected papers I ventured to describe him as my patron Saint as regards the papers contained in that volume. It brings vividly home to me how great a man he was when I reflect that to one incompetent to appreciate fully one half of his work yet he appears as a star of the first magnitude.

It affords an interesting study to attempt to analyze the difference in the textures of the minds of pure and applied mathematicians. I think that I shall not be doing wrong to the reputation of the psychologists of half a century ago when I say that they thought that when they had successfully analyzed the way in which their own minds work they had solved the problem before them. But it was Sir Francis Galton who shewed that such a view is erroneous. He pointed out that for many men visual images form the most potent apparatus of thought, but that for others this is not the case. Such visual images are often quaint and illogical, being probably often founded on infantile impressions, but they form the wheels of the clockwork of many minds. The pure geometrician must be a man who is endowed with great powers of visualisation, and this view is confirmed by my recollection of the difficulty of attaining to clear conceptions of the geometry of space until practice in the art of visualisation had enabled one to picture clearly the relationship of lines and surfaces to one another. The pure analyst probably relies far less on visual images, or at least his pictures are not of a geometrical character. I suspect that the mathematician will drift naturally to one branch or another of our science according to the texture of his mind and the nature of the mechanism by which he works.

I wish Galton, who died but recently, could have been here to collect from the great mathematicians now assembled an introspective account of the way in which their minds work. One would like to know whether students of the theory of groups picture to themselves little groups of dots; or are they sheep grazing in a field? Do those who work at the theory of numbers associate colour, or good or bad characters with the lower ordinal numbers, and what are the shapes of the curves in which the successive numbers are arranged? What I have just said will appear pure nonsense to some in this room, others will be recalling what they see, and perhaps some will now for the first time be conscious of their own visual images.

The minds of pure and applied mathematicians probably also tend to differ from one another in the sense of aesthetic beauty. Poincaré has well remarked in his *Science et Méthode* (p. 57):

"On peut s'étonner de voir invoquer la sensibilité à propos de démonstrations mathématiques qui, semble-t-il, ne peuvent intéresser que l'intelligence. Ce serait oublier le sentiment de la beauté mathématique, de l'harmonie des nombres et des formes, de l'élégance géometrique. C'est un vrai sentiment esthétique que tous les vrais mathématiciens connaissent. Et c'est bien là de la sensibilité."

And again he writes:

"Les combinaisons utiles, ce sont précisément les plus belles, je veux dire celles qui peuvent le mieux charmer cette sensibilité spéciale que tous les mathématiciens connaissent, mais que les profanes ignorent au point qu'ils sont souvent tentés d'en sourire."

Of course there is every gradation from one class of mind to the other, and in some the aesthetic sense is dominant and in others subordinate.

In this connection I would remark on the extraordinary psychological interest of Poincaré's account, in the chapter from which I have already quoted, of the manner in which he proceeded in attacking a mathematical problem. He describes the unconscious working of the mind, so that his conclusions appeared to his conscious self as revelations from another world. I suspect that we have all been aware of something of the same sort, and like Poincaré have also found that the revelations were not always to be trusted.

Both the pure and the applied mathematician are in search of truth, but the former seeks truth in itself and the latter truths about the universe in which we live. To some men abstract truth has the greater charm, to others the interest in our universe is dominant. In both fields there is room for indefinite advance; but while in pure mathematics every new discovery is a gain, in applied mathematics it is not always easy to find the direction in which progress can be made, because the selection of the conditions essential to the problem presents a preliminary task, and afterwards there arise the purely mathematical difficulties. Thus it appears to me at least, that it is easier to find a field for advantageous research in pure than in applied mathematics. Of course if we regard an investigation in applied mathematics as an exercise in analysis, the correct selection of the essential conditions is immaterial; but if the choice has been wrong the results lose almost all their interest. I may illustrate what I mean by reference to Lord Kelvin's celebrated investigation as to the cooling of the earth. He was not and could not be aware of the radio-activity of the materials of which the earth is formed, and I think it is now generally acknowledged that the conclusions which he deduced as to the age of the earth cannot be maintained; yet the mathematical investigation remains intact.

The appropriate formulation of the problem to be solved is one of the greatest difficulties which beset the applied mathematician, and when he has attained to a true insight but too often there remains the fact that his problem is beyond the reach of mathematical solution. To the layman the problem of the three bodies seems so simple that he is surprised to learn that it cannot be solved completely, and yet we know what prodigies of mathematical skill have been bestowed on it. My own work on the subject cannot be said to involve any such skill at all, unless indeed you describe as skill the procedure of a housebreaker who blows in a safe-door with dynamite instead of picking the lock. It is thus by brute force that this tantalising problem has been compelled to give up some few of its secrets, and great as has been the labour involved I think it has been worth while. Perhaps this work too has done something to encourage others such as Störmer* to similar tasks as in the computation of the orbits of electrons in the neighbourhood of the earth, thus affording an explanation of some of the phenomena of the aurora borealis. To put at their lowest the claims of this clumsy method, which may almost excite the derision of the pure mathematician, it has served to throw light on the celebrated generalisations of Hill and Poincaré.

I appeal then for mercy to the applied mathematician and would ask you to consider in a kindly spirit the difficulties under which he labours. If our methods are often wanting in elegance and do but little to satisfy that aesthetic sense of which I spoke before, yet they are honest attempts to unravel the secrets of the universe in which we live.

We are met here to consider mathematical science in all its branches. Specialisation has become a necessity of modern work and the intercourse which will take place between us in the course of this week will serve to promote some measure of comprehension of the work which is being carried on in other fields than our own. The papers and lectures which you will hear will serve towards this end, but perhaps the personal conversations outside the regular meetings may prove even more useful.

* *Videnskabs Selskab*, Christiania, 1904.

INDEX TO VOLUME V

Cambridge:

PRINTED BY JOHN CLAY, M.A.

AT THE UNIVERSITY PRESS

Printed in the United States
By Bookmasters